A history of nature conservation in Britain

'Evans provides a wealth of information on virtually every aspect of the history of nature and landscape conservation in Britain from the early beginnings to the threshold of the twenty-first century. Written from his own experience and extensive knowledge, this fully revised second edition will both enthuse and inform all those concerned with environmental and countryside issues.'

Cynthia Blackwell, *Huddersfield University*

Our attitudes towards 'nature' and the countryside are fickle. The conservation movement, despite enjoying its highest membership ever, has achieved only limited success over the last one hundred years of campaigning. Can conservationists now shake off their insular, disunited and negative image so as to gain the influence that the size of their movement warrants?

A History of Nature Conservation in Britain traces the rise of the conservation movement from its beginnings in Victorian coffee-houses to today's societies with their membership numbering in the millions. The first complete history of the British – and oldest – branch of the movement, David Evans's book offers invaluable insights into the campaigns for countryside protection and access – from battles against the use of pesticides, against pollution and genetic engineering through to legislation for the protection of our wildlife and the freedom to walk the mountains.

This second edition has been fully revised and updated. Topical issues are considered afresh; and new chapters reflect the rapid changes throughout the 1990s both in social attitudes, conservation practices, legislation, funding and within conservation organizations themselves. In the light of recent developments, Evans also looks at some difficult choices to be made in years ahead and asks how the conservation movement will fare on the new global stage.

David Evans is a Sergeant in the Gloucestershire Constabulary and has been active in nature conservation for more than thirty years. During this time, he has both worked on practical projects and run field groups on a number of reserves throughout Britain.

A history of nature conservation in Britain

Second edition

David Evans

London and New York

First published 1992
by Routledge
11 New Fetter Lane, London EC4P 4EE
29 West 35th Street, New York, NY 10001

Second revised and enlarged edition published 1997
by Routledge
11 New Fetter Lane, London EC4P 4EE

Simultaneously published in the USA and Canada
by Routledge
29 West 35th Street, New York, NY 10001

© 1992, 1997 David Evans

Typeset in Times by
Florencetype Ltd, Stoodleigh, Devon

Printed and bound in Great Britain by
Biddles Ltd, Guildford and King's Lynn

British Library Cataloguing in Publication Data
A catalogue record for this book is available from the British Library.

Library of Congress Cataloging in Publication Data
Evans, David
    A history of nature conservation in Britain/David Evans – 2nd ed.
    p. cm
    Includes bibliographical references and index.
    1. Nature conservation–Great Britain–History. I. Title.
    QH77.G7E9   1996

    333.95'16'0941–dc20   96–7573

ISBN 0–415–14491–4 (hbk)
ISBN 0–415–14492–2 (pbk)

To
Julia
John, Vikki, Joanna

If one day the cities disappear
The fields will survive.
But if the fields disappear
The cities will not survive.

Abraham Lincoln

In the environment,
Every victory is temporary,
Every defeat permanent ...

Thomas Jefferson

Contents

Plates

Figures

Tables

Acknowledgements

My thanks go to the staff of the various organisations and agencies who readily provided information and comment, sometimes at short notice. Particular appreciation goes to Miriam Angier and Deborah Lass at the Countryside Commission for allowing me access to an impressive selection of photographs and to Tim Walker and the staff at Gloucestershire Constabulary's Photographic Department for producing the copy prints. I must also commend the Gloucestershire Library Service, especially the ladies at Dursley library, for their willing help in tracking down so many titles.

At Routledge, I am indebted to all those involved in this project; in particular, to Tristan Palmer for his interest and enthusiasm for a first edition from an unknown author and to Sarah Lloyd and Matthew Smith for encouragement and advice with the second edition.

I dedicate the book to my family for putting up with it when I appeared to be dedicating myself to the book.

I take responsibility for any errors.

Abbreviations

Organisations and agencies

ABWW	Association of Bird Watchers and Wardens
AES	Amateur Entomological Society
ANP	Association of National Parks
APRS	Association for the Protection of Rural Scotland
BANC	British Association of Nature Conservationists
BASC	British Association for Shooting and Conservation (and see WAGBI)
BBCS	British Butterfly Conservation Society (now Butterfly Conservation)
BES	British Ecological Society
BFSS	British Field Sports Society
BHS	British Herpetological Society
BNA	British Naturalists' Association
BOC	British Ornithologists' Club
BOU	British Ornithologists' Union
BSBI	Botanical Society of the British Isles
BTCV	British Trust for Conservation Volunteers
BTO	British Trust for Ornithology
CARE	Community Action in the Rural Environment
CC	Countryside Commission (England and, until 1991, Wales)
CCS	Countryside Commission for Scotland (now part of SNH)
CCW	Countryside Council for Wales
CEED	UK Centre for Economic and Environmental Development
CLA	Country Landowners' Association
CNP	Council for National Parks
CoEnCo	Council for Environmental Conservation
CPRE	Council for the Protection of Rural England
CPRW	Council for the Protection of Rural Wales

CPS	Commons Preservation Society (and see OSS)
DAFS	Department of Agriculture and Fisheries in Scotland
DoE	Department of the Environment
EEB	European Environmental Bureau
EN	English Nature
FC	Forestry Commission
FFPS	Fauna and Flora Preservation Society (now Fauna and Flora International)
FoE	Friends of the Earth
FSC	Field Studies Council
FWAG	Farming and Wildlife Advisory Group
HMIP	HM Inspectorate of Pollution
ICBP	International Council for Bird Preservation (now BirdLife International)
ICCE	International Centre for Conservation Education
INCA	Industry and Nature Conservation Association
IUCN	International Union for Conservation of Nature and Natural Resources (now IUCN – The World Conservation Union)
IUPN	International Union for the Protection of Nature (and see IUCN)
IWRB	International Waterfowl and Wetlands Research Bureau (now Wetlands International)
JCCBI	Joint Committee for the Conservation of British Insects
JNCC	Joint Nature Conservation Committee
MAFF	Ministry of Agriculture, Fisheries and Food
MCS	Marine Conservation Society
NC/NCC	Nature Conservancy / Nature Conservancy Council (now EN)
NERC	Natural Environment Research Council
NFU	National Farmers' Union
NRA	National Rivers Authority
NRIC	Nature Reserves Investigation Committee
NT	National Trust
NTS	National Trust for Scotland
OSS	Open Spaces Society (and see CPS)
RES	Royal Entomological Society
RSNC	Royal Society for Nature Conservation (and see SPNR) (now the Wildlife Trusts Partnership)
RSPB	Royal Society for the Protection of Birds
RSPCA	Royal Society for the Prevention of Cruelty to Animals
SEPA	Scottish Environmental Protection Agency
SERA	Socialist Environment and Resources Association
SNH	Scottish Natural Heritage
SPNR/	Society for the Promotion of Nature Reserves/

SPNC	Conservation (and see RSNC)
TRUE	Trust for Urban Ecology
USPC	Ulster Society for the Preservation of the Countryside
WAGBI	Wildfowlers' Association of Great Britain & Ireland (and see BASC)
WWF	World Wildlife Fund / World Wide Fund for Nature
WWT	Wildfowl and Wetlands Trust
YHA	Youth Hostels Association

Designations

AGLV	Area of Great Landscape Value
AONB	Area of Outstanding Natural Beauty
ASSI	Area of Special Scientific Interest (in N. Ireland)
ESA	Environmentally Sensitive Area
LNR	Local Nature Reserve
MNR	Marine Nature Reserve
NNR	National Nature Reserve
NSA	National Scenic Area
SAC	Special Area of Conservation
SPA	Special Protection Area
SSSI	Site of Special Scientific Interest

Orders, campaigns, etc.

CAP	Common Agricultural Policy
CITES	Convention on International Trade in Endangered Species
ECY	European Conservation Year 1970
EE	Ecosystem Evaluation
EIA	Environmental Impact Assessment
ENCY	European Nature Conservation Year 1995
EYE	European Year of the Environment 1987–8
IPC	Integrated Pollution Control
LASDO	Landscape Areas Special Development Order
LCO	Landscape Conservation Order
LIFE	European Union Financial Instrument on the Environment
NCO	Nature Conservation Order
PDO	Potentially Damaging Operation
UNCED	United Nations Conference on Environment and Development
UNEP	United Nations Environment Programme
WCS	World Conservation Strategy

Introduction

Britain is widely accepted as having the most comprehensive and the most advanced system of nature conservation in the world. And that, indeed, is the fact. In no other country in the world is there so comprehensive a network. In no other country in the world has the cause of conservation so widespread, and indeed so passionate, a measure of public support.

(Vesey-Fitzgerald 1969)

The nature conservation movement as we know it in Britain today is one hundred years old. It boasts a combined membership in excess of 4 million. The National Trust 'for places of historic interest or natural beauty' is the world's largest conservation society, with 240,000 ha of land, 900 km of coastline and a membership approaching 2.5 million. The Royal Society for the Protection of Birds, concerned solely with nature conservation, has 900,000 members, a thriving junior section and almost 100,000 ha of land. Local trusts for nature conservation are supported by over 250,000 members and maintain more than 2,000 reserves between them. The network of 330 National Nature Reserves (NNRs) throughout Britain reached almost 200,000 ha in 1995. There are 6,700 Sites of Special Scientific Interest (SSSIs) that extend to 1.5 million ha. Twenty per cent of agricultural land in Scotland and Northern Ireland is classed as environmentally sensitive. Fifteen per cent of England is officially of outstanding natural beauty and one-third of its coastline is recognised as Heritage Coast. Annually, over a quarter of a million volunteers spend in excess of 1.6 million days on practical conservation tasks. There are more species of bird breeding now in Britain than there were in 1900.

Britain has ... won the esteem of the international conservation community. This country is now a world leader in conservation technology, and its work on site evaluation and safeguard, reserve management, pesticides and wildlife issues, bird protection and public relations is held in high regard by other nations.

(Nature Conservancy Council 1984)

But let us now put that into perspective. In the same report, the official conservation agency stated:

> while Britain still has a high standing overseas in the technology of nature conservation, its international reputation as a practitioner has slipped significantly in recent years.

With a total population of 56 million, Britain can claim that no more than 7 per cent of its citizens actively support the conservation bodies. For every fourteen people, only one might belong to such a society. In fact, the percentage is lower, for many members belong to more than one group. As the 1990s opened, the voluntary bodies struggled to maintain membership and resources. The British have proved themselves less ready than most to give money or effort to conservation work – or even to voice support. Complacency is rife. In 1992, only 2 per cent of Britons put the environment top of their concerns, even though 80 per cent realised that the well-being of future generations was at risk. Only 19 per cent rated the global environment as acceptable, as against 70 per cent when it came to local surroundings. Were they afraid of being expected to do something? In its level of concern for the environment, Britain foundered seventeenth out of twenty-two nations.

Total land area approximates 22.7 million ha. All the designated National Parks, nature reserves and country parks extend to no more than 10 per cent of this total area and many of them overlap one another. Recreation takes precedence over nature conservation on most of this land and designation means very little in the way of real protection. We give up 11,000 ha of rural land every year. NNRs cover only two-thirds the area of our private gardens. In contrast, agriculture accounts for 77 per cent of Britain, forestry 10 per cent and urban land use about 10 per cent. Wildlife habitat and species continue to disappear at their hands. The much heralded Wildlife and Countryside Act of 1981, which sought to protect habitats and rare species, failed to do so; creatures such as the natterjack toad and smooth snake continue their headlong decline. In the period 1970–90, nineteen species of bird declined by at least a half, including once familiar sights like the skylark (54 per cent) and the song thrush (64 per cent). Annual hedgerow loss through uprooting and neglect still runs at over 18,000 km – the distance from Britain to New Zealand. Four thousand ponds disappear each year. Leases on many NNRs are running out and SSSIs are routinely damaged by the hundred – and 80 per cent of the damage is legal.

There is no money for conservation and amenity. Funding for the official agencies levelled out and began to fall in the 1990s. Grant-in-aid to the four bodies (see Fig. 10.2) amounted to less than £150 million in 1994–5. This compared with £2,810 million for MAFF, £1,980 million for the Department of the 'Environment' and £5,850 million for the Department of Transport. Political apathy extended to a perfunctory attitude towards designated areas of supposed protection and a general ignorance of the basic principles of nature conservation.

So how effective has the movement become in its first one hundred years?

This book charts the history of the movement from the earliest days of coffee-house social clubs to contemporary societies that can boast memberships in the millions. Running alongside it is the story of the movement's campaign for access to the countryside and for positive action to conserve the environment. It is a campaign that has achieved only limited success.

We find that the conservationist ethic has been based more on sentiment than on science and that personal attitudes to 'nature' and the country-side are fickle, to say the least. Yet, for all its failings, the movement has shown continuing growth.

We look at the directions in which recent developments are taking us and at some of the unwelcome choices for the future and we wonder if the new global concerns might invest the movement with an increase in influence to match its increase in size during its first century.

1 The why and the wherefore

The facts and figures that make up the landscape of Britain today pose a pertinent question: has 'the most comprehensive and the most advanced system of nature conservation in the world' had the right effect? Has it done anything towards conserving nature or has it merely built itself into an impressive but toothless piece of machinery? Should we look to the figures with pride and satisfaction or do the facts warrant disappointment and concern?

Quite clearly, the movement in itself is very healthy indeed. The voluntary bodies continue to grow in strength and number, part cause and part effect of an increasing interest in conservation generally. But fundamental changes in outlook are needed before the conservationists can expect to succeed on a scale that befits their numbers.

The history of the movement shows it always to have been fragmented and parochial; very often, not a movement at all. Groups have studied their own speciality in isolation from all other interests both within and without the sphere of nature conservation. It is astounding that many groups still work in that fashion, even though we now accept the interconnection of all interests through the 'web of life' – all creatures and plants depending, in some way, on all the rest.

How can these groups expect to survive as responsible societies? How have they survived in the past? How, in fact, has the 'movement' come through its first hundred years?

It has survived thanks to two characteristics which, at first sight, appear to be weaknesses but which have in reality ensured its success and should continue to do so if handled wisely. They have no firm scientific base, still less an economic one. The first is the aesthetic appeal of the natural world. The second is the peripheral importance of 'nature' to so many other interests.

AESTHETICS

Take so much time leaning on a gate and watching cattle grazing; lie down for a while and smell the good earth and the growing grass; stand for a few moments on a high down and gaze your fill on the patchwork quilt of England; visit the cloistered peace of a large wood in which the noise and bustle of your daily life will seem very far away; add just a small bunch of wild flowers to taste; and be grateful to the feathered choir around you for their melody. And there you have it, a country mixture which will do the winter-worn town dweller more good than gallons of medicine.

(A.G. Street, *A Country Calendar*)

Any number of passages with a similar sentiment can be found in English literature. Reasons for conserving nature are many, varied and, most significantly, often very personal. Ask any conservationist to disregard the well-rehearsed arguments and you will uncover a deep and personal reason; the one that makes it the right thing to do; the one that makes them conservationist in outlook. They may struggle to find the words but, even without them, the tune rings true within.

much of what we call beauty in the countryside stems from conditions of ecological repose. We sense the beauty even if we have not a clue about the ecology.

(Darling 1971)

Understanding nature and acting accordingly safeguards a certain set of physical human needs. But there are other needs, less tangible and often less articulatable, yet long characteristic of our species. These involve the exercise of sentiment and the stimulation of our sense of order and purpose, our appreciation of beauty and freedom.

(Durrell 1986)

It has been said that 'as an argument on its own, beauty – natural or aesthetic – has never cut much ice in England' (Bonham-Carter 1971). In official circles, that may be the case, but officialdom does not indulge in sentiment. Beauty cut ice with the founders of the National Trust and it cuts ice with millions of concerned people today. This accounts for the movement's reliance, past and present, on voluntary effort and why so many people who do not count themselves as naturalists in any way support the conservation organisations. It also accounts for our 'potentially dangerous preoccupation with the exotic and the rare' (Mabey 1980). Many conservationists look upon efforts to save near-extinct species, or those close to the edge of their natural range, as futile and expensive exercises.

Let us face the issue fairly and squarely. A large number of animals and plants now have little or no hope of survival in the wild. It is an inevitable part of the whole process of evolution that extinction overtakes those which have failed to adapt themselves to changing environments and circumstances. Also many of the niches, which would have offered chances of survival for at least a long time ahead, have been destroyed, albeit unwittingly by man.

(Stamp 1969)

But for personal and aesthetic reasons, we continue to do our best for such species.

It is perhaps traditional that the people of Britain, sometimes slow to accept an idea, once they are convinced, take up a cause with widespread and tenacious enthusiasm. If the cause is sufficiently worth-while, there is always a vast reserve of voluntary effort. No one expects to be paid for doing a job which obviously must be done.

(Stamp 1969)

There is also the more negative but equally valuable and valid reason: the fight against something evil or ominous; and here, if ever, is an example of the 'tenacious enthusiasm' referred to above.

So there are dogged defensive actions, backs to the wall, adrenalin constantly flowing, time and money sacrificed, relations grieved, acquaintances lost; but it is enormously worthwhile; all the time there is the deep satisfaction of knowing that this is a battle not solely to save a last wilderness and the living past but also against greed and materialism. Few causes could be more inspiring.

(Sayer, in Lowenthal and Binney 1981)

The movement relies upon its inspired and committed practitioners, whether their reasons be negative or positive, their actions defensive or combative. While the cause remains to them 'enormously worthwhile', all will not be lost. But for each of them, the inspiration issues from a different sentiment; the satisfaction settles on a different set of values. We can only ask each in turn and hope to be allowed a glimpse of what makes them tick.

I know few pleasures greater than standing in a newly established nature reserve which one has helped to set up. You feel that you have made a special link with the plants and animals around you and with the people who will come to look at their descendants in the future. It is difficult to analyse why this is so satisfying. It is not just because a conservation battle has been won. I suspect that the underlying pleasure – relief is almost a better word – is connected with a desire to produce something permanent in today's world, in which so much else is subject to unpredictable change. Setting up a nature reserve has similarities with painting a picture or writing a scientific paper: a great deal of hard work has been done, obstacles have been overcome and, at the end, something new has been created. One has impinged on the future as well as the present. Of course, nothing can be predicted with absolute certainty – political madness or war may undo one's

work or mismanagement may mar it – but what can be done has been
done.

(Moore 1987)

Feelings; personal conviction. The cause of nature conservation – any
cause – could have no stronger ally. The things closest to a person's heart
are the things for which that person will fight most keenly. The voluntary
organisations have always used this fact most wisely to attract support.
Their advertising focuses upon suffering or loss, on damage or destruction
of those things dearest to us. Thus we are shown poisoned animals, trapped
birds, toppled trees, trodden plants, bland or barren landscapes. 'Direct
action' groups, most notably Friends of the Earth and Greenpeace, have
attracted much public sympathy for the cause, in spite of some extravagant tactics, by highlighting emotive issues such as the culling of baby
seals and the dumping of noxious chemicals in our beloved seas. Even
efforts on behalf of our less-loved creatures have played upon emotions,
such as the campaigns of the Fauna and Flora Preservation Society on
behalf of bats and snakes.

But even feelings have not cut much ice in official circles. Such dearly
held personal conviction ought to have done but, because it cannot be measured and evaluated, and because the conservationists have not spread the
word sufficiently, it has been ignored by the policy-makers. It is these two
extremes of reaction to personal emotions that account for the movement's
having become so large while its practical results have remained so slight.

> Feelings are shunned in the sharp scientific discourse of modern ecolog-
> ical conservation – which I find rather sad, because a compassion for
> and a delight in the natural world are what turn people to act in its
> defence in the first place. The fact that such feelings are impossible to
> quantify does not make them irrelevant.

(Mabey 1980)

On the contrary, because they cannot be quantified, it is all the more
important to give them full consideration. But in an economic world, if a
thing cannot be costed, it remains irrelevant. In a legal world, feelings
have no sway. In a geographical world, climate and topography dictate
the best land use. Only in a political world can the feelings of individuals
have any effect at all and only then if those individuals present a sufficiently united front. This is the main hope for the nature conservation
movement and very gradually it is coming to realise this fact.

Feelings, though, while bringing personal conviction to the cause, have
also been responsible for preventing the crystallisation of a 'movement'.
The 'tune within' has often been too individual, too intimate. This is where
sentiment, in one way the saviour of nature conservation, has also come
to threaten its effectiveness. Darling gave the warning in the 1969 Reith
Lecture which did so much to bring the cause to general notice.

Sentiment and ethics should never be confused. Ethics stand firm and are to be sought by spirited and intellectual effort of reflection; sentiment is a poor guide in the mosaic of ecology and conservation.

(Darling 1971)

He talks of a 'hardheaded' attitude towards conservation; specifically, to draw some monetary return from it in the then-forthcoming economic 1970s. He was attempting to focus the various attentions of the conservationists and to make them credible to those outside the 'movement'. He could see the cause and the effects of the parochial past. The cause – those personal feelings giving a blinkered and self-centred view. The effects – its two great weaknesses: its inability to unite within itself and its unwillingness to make allowance for other interests outside natural history. How could those 'outsiders' have been expected to understand the conservationist point-of-view? There was little common ground.

ASSOCIATIONS

What common ground there was arose from that second fortunate characteristic of 'nature': its peripheral but pertinent relevance to other interests. Basically, almost any activity which takes place out of doors – and many which do not – inevitably come into contact with the natural world. The huntsman has a vested interest in preserving the stocks of his quarry and, for the last thousand years, has taken positive steps to do so. Strange then that we should still see eyebrows raised when the sportsman labels himself 'conservationist'. Farming and woodland management have relied through the centuries on healthy and sustaining practices which ensured a varied flora, fauna and landscape. Variety is health in the natural world. Only in the last fifty years have the practitioners, now known as agribusinessmen and blanket foresters, taken to bulldozing nature as though it is no longer of relevance to them.

The estate owners of the eighteenth century planted trees and encouraged wildlife – apart from certain outlawed species – in order to enhance their domestic surroundings. Their leisure time gave them an interest in nature study, but they gave no thought to its conservation. Today, more people enjoy increasing hours of leisure time. Coupled with modern mobility, this has brought about an explosion of hobbies, pastimes and interests, many of which depend upon access to the countryside. The pathway to free access has always been strewn with obstacles and many remain today. Improvements have given us the National Parks and the long-distance footpaths. But access to the countryside conflicts with the basic needs of nature conservation, and the two interests have developed separately. Thus we have had a Countryside Commission and a Nature Conservancy; a National Parks system and a series of National Nature Reserves. But the world became too small to bear their conflict and the

1990s saw them accept an inevitable coming together (p. 211). Natural history is such an important feature of the countryside to which access is demanded, and freedom to get out and enjoy the countryside so vital to an interest in conservation, that the two could remain apart no longer.

Familiarity breeds, not contempt, but comprehension. Modern media have nurtured some knowledge and understanding of the countryside, but unless it is followed up with an enquiring visit, it remains a strangely isolated experience. Nature needs to be enjoyed at first hand. Leisure and mobility have done much more than any number of magazines and television programmes to cancel out the effects of enforced isolation from our countryside. Britain was the first industrialised country. Most people live in towns and have traditionally enjoyed no real contact with the countryside, so that, as Bonham-Carter (1971) has written:

> it is all too easy to convey the impression that amenity is an extra, something which – though good to strive for – can be added to life on top of the real business of existence. This attitude, typical of urbanised countries and particularly true of this one, is a heritage of the Industrial Revolution, which dehumanised work and divorced man from Nature.
> ... once the damage done by industrialism had gone deep and man had become disorientated – say by the third quarter of the 19th century – the approach (or return) to amenity was ... hesitant and piecemeal ... The whole movement, if such it can be called, was a series of actions and reactions, irregular in direction and pace, touching a variety of apparently disconnected subjects, from the protection of wild life to the betterment of housing.

The history of nature conservation in Britain has been moulded by the sportsman and the rambler; the day tripper and the country dweller; the town and country planner and, occasionally, the conservationist. Sometimes they have done good for the cause; sometimes harm, but always they have their own interests at heart. Human nature is too selfish to conserve the requirements of other living things as an end in itself. That would be looked upon as just too sentimental. We conserve for our own private reasons. No honest conservationist can deny it any more than the huntsman can. More common ground must be found between all these personal interests.

SCIENCE AND ECONOMICS

Only in the mid-twentieth century did scientific arguments come to the aid of nature conservation, providing for some a foundation for their personal reasons, for others a barricade behind which to screen them. Only in the 1980s were economic approaches made to the cause.

Science began its advance during the war years of the 1940s. At the start of the decade, Fisher (1940) wrote:

The aesthetic argument is the stimulus behind practically all the bird protection that has been organised by law and active deed in this country. Among the many facets of this argument, a new one is becoming important – that birds should be preserved, not only because of their rarity, curiosity, wildness, or romance but also because of their biological interest. This new argument contradicts some of the old ones, for many of the predatory birds disliked by sportsmen and sentimentalists are of primary interest to biologists. The new biological conservation desires very little more than the effective removal of man, alone among the predators, from a living association of birds.

Nature was now of 'biological interest'. The sentimentalists must stop disliking the predatory birds for the scientist now found them of use. Could it be that they held an important place in the natural world? Was it right that persecution of them upset a certain natural balance? A more scientific attitude to nature began to creep in. Haphazard and uncoordinated recording gave way to systematic study. Greater care was taken in the selection of nature reserves, and management plans were introduced where, earlier, nature had been left to run riot. The Field Studies Council and the Nature Conservancy were established to encourage a scientific approach to nature study. National Nature Reserves and Sites of Special Scientific Interest were identified as living laboratories. Conservation came a long way behind scientific study in post-war Britain.

But very soon the studies began to bring home two particular truths. The first was that 'nature is indivisible' (Vesey-Fitzgerald 1969). The web of life was indeed closely woven and any tear in its fabric would weaken the strands around it. That notion should have been a comfortable one: the human race, full of health and vitality, buoyed up by the inherent goodness of everything around it. But it has never been a rosy notion because it makes us dependent – dependent on almost everything in the natural world. While the scientists were seeking general acceptance for the idea of their 'web', they were also introducing the other home truth – that human activity was beginning to wear through the fabric in certain places. Much, much more work was needed to discover the full extent of the problem. But the web of life could by the same token become a web of death or, at least, of disease. The conservationist cause had found another platform.

It came to a head in the 'Doomsday' preachings of the 1960s. Much was hysterical overstatement; much was based on wayward assumption, but it focused attention on the fact that we were rapidly outgrowing the Earth. In a speech just 37 days before the end of the decade, HRH Prince Philip (1978) summed up the findings:

I realise that there are any number of vital causes to be fought for, I sympathise with people who work up a passionate concern about the all too many examples of inhumanity, injustice and unfairness, but behind all this hangs a really deadly cloud. Still largely unnoticed and unrecognised, the process of destroying our natural environment is gathering speed and momentum. If we fail to cope with this challenge, all other problems will pale into insignificance.

Science was now in the front line, supporting the aesthetic and personal approach. The main enemy now was the economic one. After another decade of falling back under the weight of vested interests, a battle was won with the publication of the World Conservation Strategy. Economics were at last on the agenda. The Strategy suffered a universally inauspicious start but its guidelines remained for all to act upon if they wished. For, if economic values could be brought into line with conservationist thinking, the cause would win its strongest ally ever and the appeal of conservation might be broad enough to interest the politicians and policy-makers. Only time would tell.

DEFINITIONS

Having looked at some of the more articulatable reasons for conservation, we must now attempt to define it. 'Many people will wonder just what this word CONSERVATION means,' mused a Nature Conservancy leaflet of the late 1960s. This was partly a result of the insular attitude of the movement up to that point, and the suggestion sadly remains valid to this day. The multiplicity of personal reasons leaves us with as many definitions as we have conservationists.

Put simply by Nicholson (1987), 'preservation' implies keeping things as they are, and 'protection' implies keeping outside interference at bay. 'Reservation', involving refuges, reserves and sanctuaries, aims to avoid use or exploitation of an area, and 'zoning' or 'segregation' seeks multi-purpose use by careful management of resources. 'Conservation in its modern sense is a broader concept, comprehending all of these limited approaches and more.'

The definitions can be put into four broad categories based on the reasoning behind conservation. The aesthetic and the scientific are, as we have seen, the best established and hence easiest to find. The economic interpretation is altogether less common, while the political angle was unheard of before the 1990s. It becomes clear by looking at various definitions that we seek to achieve three things: to maintain our quality of life (in aesthetic, not economic terms); to limit the damage we do to our planet (broadly a scientific approach); and to improve our standards of living (broadly economic). The correlations are clouded only by the inevitable overlapping of ideals, and by the political tightroping of the 1990s which tries to pay homage to the scientific without compromising the economic.

Thus the aesthetic angle:

Conservation means all that man thinks and does to soften his impact upon his natural environment and to satisfy all his own true needs while enabling that environment to continue in healthy working order.

(Nicholson 1987)

Conservation therefore means the process of reconciliation between the things which are needed for the practical satisfaction of people and those things which make life worthwhile.

(HRH Prince Philip 1978)

Health is the capacity of the land for self-renewal. Conservation is our effort to understand and preserve this capacity.

(Leopold 1949)

These commentators consider that quality of life not only depends upon, but is, a healthy environment. The same sentiment merges with the scientific basis in Darling's 1971 definition:

The art of conservation stems from the science of ecology, a delight in knowing how nature works and a love of beauty which may or may not be conscious.

Stamp (1969) was able to give a purely scientific angle:

Nature Conservation is essentially applied ecology and ecology is the study of plants, animals (and man) in relation to the environment.

But such interpretations are cold, impersonal and do little to fire the imagination. In order to win new converts, the movement needs to keep conservation meaningful.

it means caring for and making wise use of our land and water and of their plant and animal life, with the aid of scientific research and management. Conservation goes beyond protection and preservation. It involves trying to understand how nature works and using this growing knowledge to help us to work with instead of against nature.

(Nature Conservancy: undated)

The creation of a satisfactory state of existence for all living things on this earth. It includes the protection of the best of what we have inherited, the correction of the worst mistakes, and the considerate planning of future development.

(HRH Prince Philip 1978)

In so far as damage is inevitable to human enjoyment it must be restricted to the income of life and not infringe upon its capital.

(National Trust 1986)

Thus the scientific argument begins to take on an economic flavour in an effort to make itself meaningful in the modern world. For so long, conservationists were seen as anti-progress and, as a result, the barriers went up. The label has not yet been removed completely, but conservation is at last relating to economics through its undoubted connections. The MacEwens (1982) sum it up as:

> the thrifty use of non-renewable resources and the use of renewable resources without diminishing their quality or endangering their supply.

But the interpretation needs to go further than this. Conservation is the way not only to high quality of life but also to a sustainable high material standard of living:

> the management (which term includes survey, research, administration, preservation, utilisation, and implies education and training) of air, water, soil, minerals, and living species including man, so as to achieve the highest sustainable quality of life.
>
> (IUCN 1980)

> meeting the immensely diverse human needs from the natural world [is] the goal of nature conservation, rather than an obstruction to it.
>
> (Mabey 1980)

Such definitions may sound like capitulation to greed and materialism but in fact they waymarked a wise change of course by conservation in its quest to meet up with economics. They were not a sop but a call of concern to encourage the economist to take a more long-range view of his activities. By sensible use of all that the world provides, mankind will continue to live well and everything else on earth ought to survive.

Now the politicians felt able to join the conservation bandwagon. They could go along with the concept now that it did not appear to threaten future living standards which, in their book, of course, were monetary standards. So, when Ian Lang, Secretary of State for Scotland, introduced the idea to the House of Commons in February 1991, he said that:

> the environment should be so regarded and maintained that it does not erode or degrade and is handed on to future generations in the same condition or possibly enhanced or developed. Therefore no operation should be allowed to take place which would damage the environment without restoring or replenishing the damage.

This did not, in itself, preclude continuing development in the traditional sense; it merely called for extra environmental consideration. 'Sustainability' became the political buzzword of the 1990s. It is, in fact, conservation in the truest sense of the word:

> the use of the natural resources for the greatest good of the greatest number for the longest time.
>
> (W.J. McGee, quoted by Nicholson 1970)

That, surely, is something we all want.

The nature that we seek to conserve, then, is more than the wildlife. In reality, the basic genetic make-up of living things has only recently become a stated part of the equation. Given the variety of standpoints for conservation, it was inevitable that 'nature' should also, at times, have included varying elements of scenery, geological forms, mineral resources; everything, in fact, not only in the countryside but increasingly in the urban landscape too.

In the next chapter, we shall see that much of this 'nature' is, in fact, very unnatural. Even some of the animals and plants have been artificially introduced. We shall also see how human attitudes, upon which any movement must be founded, are wont to vary according to one's place in history and one's outlook on life. The conservationist ethic is based on very shifting sands.

2 'Prehistory': Isolation and ideals

ATTRITION

> As soon as he became man ... he began to alter the face of the natural world as it was until then. While he remained a hunter-foodgatherer he was little more than another indigenous animal. But as soon as he burned wood to keep warm he was consuming it in a different way from natural decay, with different consequences. When he used fire as an aid to driving wild animals into places where he could kill them more easily, and burned bush to encourage hoofed animals to graze on the young grass which followed, he had begun his ceaseless attrition of the natural wilderness.

> (Darling 1971)

Britain's natural wilderness, in the present interglacial climate, is one of trees: primarily oak, ash, hazel and elm, with beechwood on the chalk and limestone, maple and lime in low-lying parts of England, silver birch on higher ground. In Scotland, the birch predominates in the west while the great pinewoods of Caledonia – the first to form when the last ice receded – clothe the eastern regions. Only on the highest plateaux or in the extreme north-east, 'the flow country', does woodland naturally give way to moorland. Only here would a bird's-eye view look down upon something other than the canopy of trees.

That the birds see a different picture is due solely to the presence of the human race. *Homo sapiens* is but one of three million known species of life on earth yet we eat more food than all other land animals put together and, in 1986, food production per head of population worldwide began to decline. In hunting, rearing, cultivating and processing this food, we have momentus effect upon the natural landscape. Being successful in this has led us to increase in number and to seek more material artefacts in life. Our domination of the landscape has thus grown exponentially, creating an almost entirely artificial environment.

Until about 6,000 years ago, the wilderness remained unaltered by mankind. Towards the end of the Mesolithic era, some limited woodland clearance began. The primitive agriculture that induced this move

Plate 2.1 Left to its own devices, Britain's landscape would revert to almost total tree cover: beech trees in the Blackdown Hills AONB. Photograph by Peter Hamilton. Courtesy of Countryside Commission

expanded greatly during the ensuing Neolithic period (4,500–2,000 BC) and the first real effects of man upon British habitats and landscape were felt. Rackham (1980) points out that elms declined by 50 per cent and describes how conditions were being provided which suited the spread of elm disease. Already, our practices were having effects over and above the obvious.

Heathland, a new habitat in Britain, formed as areas such as the East Anglian Breckland were cleared of trees. Moorland spread into previously wooded regions, some of them high and difficult of access, such as the Lake District. Other areas were problematic for different reasons, such as the waterlogged Somerset Levels, and yet agriculture was not to be forestalled even then. The Bronze Age followed, by the end of which (500 BC) the nomadic way of life had been superseded by permanent settlements. About half of England had been stripped of its trees and all present British moorlands had been instigated.

The Romans left behind their towns and roads to impress us in the twentieth century, but they farmed little of Britain, leaving much of the countryside to revert, via heath and scrub, to natural woodland under their regime. Their drainage schemes had some effect upon landscape and wildlife but many soon fell into disrepair. When the Anglo-Saxons reached Britain in the fifth century, they found much derelict farmland to reinstate as well as wilderness to reduce for agriculture. Their system of open-field farming, centred upon villages, was responsible for woodland clearance over vast new areas so that by 1086, the year of Domesday, only 15 per cent of England remained wooded, moorland had increased again in extent and heathland was more widespread than at any other time before or since. Drainage schemes had been applied to large areas, including much of Somerset and Kent. Yet there was still much damage to be done, for the landscape bore no resemblance to what we see today.

> With an average of only about twenty-five people to the square mile [10 per square kilometre], England at the end of the eleventh century was greatly under-populated. Thousands of square miles were still untouched by plough or beast, thousands more only half used in a shifting cultivation of out-field. The rich natural resources of mineral wealth had hardly been scratched. Great tracts of forest, of moor and heath, marsh and fen, lay awaiting reclamation by pioneers.
>
> (Hoskins 1977)

There then followed the effects of two hundred years 'of rising population and of such land-hunger as we have never experienced since' (Rackham 1986). An average of 7 ha of English woodland per day was cleared for arable farming, reducing total cover to just 10 per cent by 1350. Agriculture moved onto the moors and further into the heaths. Intensification began and, in many regions, the landscape became an agricultural one. Not until the onset of the Black Death in 1348 did the countryside again enjoy a respite – of some 500 years – from the attentions of a devastated human population.

Inevitably, there had been great changes in Britain's native wildlife through these ages. Some were due entirely to climatic or natural habitat change, such as the loss of the reindeer soon after the ice receded. The elk had gone by the early Stone Age and the lynx by the late. The aurochs or wild ox, now extinct worldwide, died out in England and Wales about 600 BC but may have survived in Scotland until the ninth century AD.

Human presence affected many other species. The brown bear, hunted for meat and fur, was driven out in around AD 900. The beaver, killed for its pelt, was last seen in Wales in 1188. Hunting was so widespread, popular and indeed necessary for food and clothing that the Royal Forests were set up by William the Conqueror in order to reserve their excellent hunting potential for royalty and nobles. 'Forests' were, in reality, the cleared areas where hunting was at its best and the kings jealously guarded their self-assumed rights. Anybody could hunt predators such as the wolf and the wild boar, but deer poachers under William risked losing their hands. Clearly this threat was not deterrent enough, for the punishment was later increased to execution, and access to the hunting parks and forests was restricted further. The Royal Forests covered large areas, such as the Forest of Dean and the New Forest. Restricted access resulted in under-use, and inadvertently they became havens for wildlife. Natural regeneration encouraged a varied flora and fauna. 'By the end of the Middle Ages the hunting reserve was a well recognised technique for preserving stocks of game' (Fitter and Scott 1978). But any benefit to wildlife was incidental to the interests of the huntsmen, and outside the Royal Forests and parks the decimation of wildlife went on unabated.

Hunting became more effective as firearms began to replace the crossbow from the 1630s onwards. But it was the wholesale change to natural habitats that combined with hunting to bring about the most devastating and permanent effects on native wildlife. In England and Scotland, the wild boar hung on until 1676, when hunting and inbreeding with domestic animals took their final account. Yet for the last four hundred years of its existence, people in certain parts of Britain had encouraged its survival by providing food in winter. The wolf also enjoyed some protection under the Tudors, who introduced a close season for three months from Christmas Day. Having survived open hunting since the Anglo-Saxons had realised the need to protect their sheep, the wolf was now threatened by the loss of forest. The last English wolves disappeared from Yorkshire during the reign of Henry VII, but in Scotland, they survived for another two centuries at least. Some put the date there as 1680, some as 1743. In Ireland, the last known wolf was noted in 1786.

The changes in agriculture that hastened woodland clearance suited a more open type of sport, and hunting with hounds began to gain respect from the time of the reign of King John (1199–1216) onwards. Hares were the favourite quarry; deer were also popular. The fox was not widely valued as good sport but it slowly gained favour. During the eighteenth century, it came to be hunted more regularly than deer and, by the end of the nineteenth, it had ousted the hare as the favourite quarry. Otter hunting too had its origins in the days of King John and increased in popularity up to the eighteenth century. These sports still hold a place in traditional country life (Table 2.1) although changing fashions and levels of population force some of them to be suspended from time to time. If the recovery of the otter population, for instance, ever removes otters from the danger list, calls for hunting to resume are sure to be heard. The effect of hunters on the populations of their game is minimal in comparison with other factors, and any resistance to their continuance is based more on sentimental than conservationist ethics. The final demise of these sports, if it comes not as a result of other factors wiping out the quarry, will be a response to changing human sensibilities.

The other popular sport of the early centuries was falconry. So popular was it during King John's reign that he had to ban it for a season to allow stocks of game to recover. Wildfowling too was banned for a season because snaring and shooting (with crossbows!) had so depleted populations. From the twelfth to the seventeenth centuries, the interests of falconry gave some protection to many birds of prey. In Henry VII's day, the punishment for taking eggs of certain raptors was imprisonment for a year and a day. Then during the seventeenth century, falconry lost popularity to wildfowling and with it went the protective measures for birds of prey. Their numbers began to fall, hastened by the effects of enclosure and the increasing pressure to protect game species (p. 30).

But, like today, the effects of hunting were nothing compared to those of habitat loss.

> There is . . . some evidence that woodland clearance began in Mesolithic times, say seven or eight thousand years ago, though their farming could only have produced very limited clearances in the woods and all the signs have been lost in the re-invasion by trees and heaths.
>
> (Hoskins 1977)

In the fifteenth century, however, the tears and gashes in the woodland cover really began to affect the landscape. The Tudors were concerned to slow the rate of felling, though not for the sake of the wolf nor any other wildlife. Timber was a valuable resource for housing, shipbuilding and iron smelting and had been 'managed' since Norman times. The Tudors introduced Acts of Parliament to ensure good husbandry and to moderate the amount of timber being used by industry. The spruce and silver fir, the larch and holm oak were introduced from abroad. Even at that time, the needs of farming were such that the conversion of coppice woodland to arable or pasture-land had to be forbidden. When Thomas (1983) states: 'The attitude of agricultural improvers to woods and trees was

Table 2.1 Hunting packs in Britain and Ireland (1990)*

Fox		208 : England, Scotland and Wales
		34 : Ireland
Stag/Deer		4 : England
		2 : Ireland
Hare	–harriers	27 : England, Scotland and Wales
		25 : Ireland
	–beagles	81 : England, Scotland and Wales
		30 : Ireland
	–bassets	12 : England, Scotland and Wales
		1 : Ireland
Otter		4 : Ireland[†]
Mink		20+ : England and Wales

* Some packs inactive at this time.
† As recently as 1978, seventeen packs in England, Scotland and Wales, now all disbanded

usually extremely hard-nosed', he is referring, not to the 1980s, but to the sixteenth to eighteenth centuries. He finds examples, even during the height of the enclosures, of farmers who would not tolerate trees or hedgerows and quotes Joshua Poole in *The English Parnassus* of 1657, describing forests as dreadful, gloomy, wild, uncouth and melancholy. It is no wonder that the needs of industry continued to be appeased and that the new Tudor laws failed to stem forest decline. Whereas the Tudors had begun the sixteenth century with at least 1.6 million ha of woodland, it had been reduced by a quarter by the late seventeenth century. Fifty per cent of England and Wales was by then committed to agriculture. There were 1.2 million ha of artificial forests, parks and commons, which enjoyed far greater popularity and acceptance than the 4 million ha of heath, moor and mountain.

All types of natural wilderness, in fact, were regarded with scorn, while agriculture, on the other hand, was a wonderful improver of the landscape. A cultivated field or a lush pasture was a joy to the eye. Uncultivated land was a waste and a disgrace; on the same level as a forest or an open heath. In the name of agriculture, weeds were eradicated at every turn. Many animals, too, faced persecution: rats and mice, hedgehogs and moles, weasels and stoats, jays and bullfinches, crows and many birds of prey. In 1534, the declining red kite had been given protection by Henry VIII because of its value as a scavenger and cleanser of town and countryside. This was the first time that any bird had been protected for a reason other than sport. By the following century, no longer a vital part of the cleansing operation, it was being shot on a wide scale for attacking poultry and for its other deleterious effects on farming.

Wetland was being turned over to agriculture. Drainage schemes were undertaken which had enduring effects upon wildlife. The work of the

Romans had removed the Dalmatian pelican from Britain. Vermuyden, the famous Dutch engineer, drained vast areas between 1630 and 1655. Cranes and spoonbills nested in the Fens during the sixteenth century, and the same act of 1534 which protected the kite also protected, between 31 May and 1 August, cranes, spoonbills, bitterns, herons and great bustards, and their eggs. Although the loss of the crane by the start of the seventeenth century can be attributed to natural causes, the spoonbill's disappearance by the end of that century must be blamed, in part at least, on the loss of habitat through drainage.

ATTITUDES

In the light of such enthusiasm for the taming of the wilderness into preferred artificial landscapes – an enthusiasm often accredited only to the twentieth century – we must ask ourselves the question: why was it that attitudes changed so radically during the seventeenth to nineteenth centuries that the seeds of the conservation movement were able to germinate? They changed in two ways – a more caring and responsible attitude towards animals and a new appreciation of wild landscape – and for two reasons: a reappraisal of our place in the world and a new isolation from nature.

The way we treat the natural world reflects our opinion of the place we hold in the overall scheme of things. During these centuries, the scriptures were reappraised in the light of changing values. Personal opinions of 'right and wrong' have always, through sentiment, played a part in conservation as they have in religion. As early as AD 680, St Cuthbert founded a wildlife sanctuary on the Farne Islands in answer to his religious beliefs. More often, though, it was considered pagan to worship the natural phenomena of air, fire and water. Wilderness was an evil place to be shunned at all costs. The Bible gave mankind power over every living creature on earth. We had always taken this to mean that the natural world was provided solely for our use. But now, more and more people were inclining towards Pope's view that the world was not quite that simple; that human dominion over them called for more responsibility towards other creatures.

> From Nature's chain, whatever link you strike, Tenth, or ten thousandth, breaks the chain alike.
>
> (Alexander Pope, *Essay on Man*, 1732–4)

The theory of the 'chain of being' can be traced back to the Ancient Greeks, as can the doctrine of 'plenitude'. The first claims that we are but one creature in a whole series of beings that starts with God and His heavenly crew (angels) and comes down to the lowest orders of creeping insects, plants and even clay. It is, therefore, a simplistic forerunner of the 'web of life'. The second suggests that no possible form of life along the chain has been denied its existence; that there are no missing links and that even imperfections and evil have their place in God's world. In the climate of the eighteenth century, many reformers sought to throw out these 'laws' which tended to suggest that social inequality was a natural phenomenon. Pepper (1984) details their fortunes but warns against reading too much into their effect upon the modern conservation movement. What is beyond doubt, however, is that the high profile these arguments enjoyed will have encouraged many individuals to give them close thought and it was from the religion and politics of the individual mind, uncluttered by teachings and manifestos, that the conservation ethic grew. God had formed a perfect world in which every creature, every plant, no matter how lowly or noxious, held a place and fulfilled a role. Variety was vital to life; every organism had its place and none should be removed. Mankind, with his power over all living things, had the ability to ensure their continuance and success; he had it within him to leave well alone – the ability, in short, to protect and conserve. Here was a springboard, if ever there was one, for any future conservation movement.

The study of natural history became a new social pastime, particularly among the middle classes. An active interest was taken in the welfare of living creatures. Animals and birds were seen as living things with rights and needs of their own. Conversely, humans were seen as one species in a wider world with responsibilities to all God's creatures.

The new acceptance of 'animal rights' began to manifest itself in sentimentalism and in a hardening of attitudes against wanton cruelty. 'It became a mark of human sensibility to throw crumbs to wild birds in winter' (Thomas 1983). The very British habit of keeping pets began in earnest at this time, and caring observation of their antics played a large part in altering popular attitudes. Yet there also arose some opposition to the keeping of living creatures in cages.

A Robin Redbreast in a cage
Puts all Heaven in a rage.

A Dove house filled with Doves and Pigeons
Shudders hell through all its regions.

William Blake continues his *Auguries of Innocence* (1789) with attacks on all forms of cruelty – starving a dog, misusing a horse, killing a fly – and, most tellingly, he finds fault with accepted sport:

Each outcry from the hunted Hare
A fibre from the brain does tear.

The first concerted calls to ban hunting for sport were heard. These conflicts over captive creatures and blood sports are with us still. There was so much writing and discussion of our moral duties towards other animals that Thomas sees it as 'one of the most distinctive features of the late eighteenth century English middle-class culture'. But it was only middle-class.

Because it taxed the moral conscience, preyed upon sentiment and had such far-reaching potential, the new code for living gained acceptance only in a slow and piecemeal fashion. The majority in the eighteenth century still lived by the old values. For the poorer classes, survival depended on hunting and agricultural improvements brought promise of easier times. The landed gentry planted their estates, but not with wildlife in mind; its well-being was no concern of theirs and their sport must continue. Even amongst the academics, there were many who saw no relevance of wildlife to human existence. When Linnaeus's classification of the animals, which placed man amongst them, was officially accepted in England in 1760, it met with much resistance and distrust in scientific circles.

Many of those who fed the birds continued to shoot them for sport. The sparrow had yet a hundred years to face the sportsman's gun. Those who called for an end to the hunting of deer and hares kept quiet on the question of otters and foxes, which were still unpopular. Nobody sought salvation for the rats that fed in the granary or the moles that burrowed in the fields. Sentimentality reached ludicrous proportions when 'pollarding' of trees, long recognised as a good management practice, declined in the face of opposition to the mutilation!

Such misunderstandings should not be laughed off as a reflection on the infancy of nature study in the eighteenth century. To this day, these confused emotions persist, causing the contradictions in personal outlook and in conservationist argument that prove so painful so frequently. Why do they arise? Because we live in isolation from nature. The new awareness of animals and birds coincided with the growth of the urban population. The cases against hunting and other sorts of cruelty – real and imagined – were made out and heard largely in the towns. Country folk, either because or in spite of their greater comprehension of the natural world, continued much as before.

It was this new-found isolation from nature that altered attitudes towards landscape as well. Distanced from the natural wilderness, seventeenth- and eighteenth-century people found that even its 'horrid' aspects began to assume a fatal attraction.

> Even the rude Rocks, the mossy Caverns, the irregular unwrought Grotto's, and broken Falls of Waters, with all the horrid Graces of the Wilderness it-self, as representing Nature more, will be the more engaging, and appear with a Magnificence beyond the formal Mockery of princely Gardens.
>
> (Shaftesbury, *The Moralists*, 1709)

The dreadful, gloomy forests of Joshua Poole began to take on a new, refreshing hue:

At the Prospect of the wide and deep Ocean, or some huge Mountain whose Top is lost in the Sky, or of an old gloomy Forrest, are not our Minds filled with a pleasing Horror?

(George Berkeley, *Three Dialogues between Hylas and Philonous*, 1713)

Salvation for the forests began as early as 1664 when, at the behest of Charles II, John Evelyn published his *Sylva*, which reviewed the sorry state of contemporary timber resources. Although it considered forests from an economic aspect, it nevertheless awakened more fundamental feelings towards them. Through the influence of Evelyn's ideas, tree-planting 'acquired the status of national policy and of a national upper class habit' (Nicholson 1970). Private landowners were soon supplying most of the Navy's timber needs and, although it was done with money in mind, there is no doubt that, on the large estates at least, planting was planned to please the eye. The cosmetic effect on the countryside, as well as on attitudes, was quite startling in some areas. The Dukes of Atholl planted 14 million larches on their Perthshire estate between 1740 and 1830. In just thirty years prior to 1816, Thomas Johnes planted 5 million trees on the Hafod estate near Devil's Bridge. Between 1760 and 1835, no fewer than 50 million trees, mainly conifers, chestnuts and limes, were planted by private landowners in Britain. Large-scale schemes were eligible for gold and silver medals awarded by the Royal Society for the Encouragement of Arts. Many of our introduced species date from this time. Gardening with trees had become a social pastime.

Conversely, the formal gardening of the seventeenth century, charac-terised by geometric patterns, fountains and sculptures, gave way to the more natural vistas of parkland. Designers, like William Kent and Capability Brown, broke new ground with their wild and romantic land-scapes. Not that many of them were natural, of course; hills were raised and lakes excavated in earth-moving exercises that might impress today.

But the search for natural landscape was on. By the mid-eighteenth century, 'something quite new in history was making its presence known: the collective enjoyment of the scenery of nature for its own sake and in its original, unmodified condition' (Allen 1976). Mountains and moorland gradually acquired a reverence of their awesome and brooding beauty. 'Not a precipice, not a torrent, not a cliff, but is pregnant with religion and poetry', wrote Thomas Gray in 1739. The composers, too, reflected the aura of natural landscape. Early in the nineteenth century, Mendelssohn had written his Third or 'Scotch' Symphony and the Fingal's Cave Overture. The Malvern Hills influenced Elgar's work, and the Dorset heaths, beloved by Thomas Hardy, affected the music of Holst in the early 1900s.

For the time being, though, heathland continued out of favour. Wetlands were drained for farming or parkland. Not all wild scenery was as readily accepted under the new regime as the forests and the mountains. This was because forests and mountains could not be turned to the cultivation of food. Heathland and wetlands could. And still 'to eighteenth century rational man beauty was well-proportioned and cultivated land, and wilderness held no attraction' (Pepper 1984). We tend to hold the same general view today.

TOWN AND COUNTRY

Yet the attraction of natural landscape was exerting a greater pull as people left the countryside and headed for the towns. In 1800, 25 per cent of the British population lived in urban areas. The percentage had doubled during the previous century and would do so again in the next fifty years. By the end of the nineteenth century, 80 per cent would be town dwellers and the British people would have become predominantly urban once and for all. What began in the eighteenth century as a gentle shift in attitudes became a major reappraisal in the nineteenth. What had been the luxurious moralising of the middle classes began to concern the masses, though not, of course, the majority. Concern for the countryside has never been a majority cause and it is unlikely ever to become one without the rallying of peripheral vested interests. But what did become evident in the nineteenth century was a wider appreciation of the countryside and from it grew the rosy, over-optimistic view of our rural heritage with which we are saddled today. There are a number of reasons for this change in attitudes, some based on the reality of what was happening to our countryside but many arising from what people who were increasingly isolated from it believed the countryside to be.

Nature and landscape of course exist in reality, but our feelings for them, upon which the conservationist ethic is based, are extremely personal and very varied. The scientist can break them down into their basic elements but, for most of us, what we value in nature and landscape is a cultural thing: an overall impression built up through life from first-hand experience and – more prominently these days, I suspect – from the influence of others, received through various media. And so it is that the conservationist mood is as much a moral or social phenomenon as it is a scientific one. In fact, while the scientific approach has recently come to win converts to the cause, through talk of global threats and the like, it is the aesthetic argument that continues to power the movement as it has always done.

Williams (1973), in fact, considers that the town versus country pretence serves only 'to promote superficial comparisons and to prevent real ones . . . to ratify an unresolved division and conflict of impulses, which it might be better to face on its own terms'. He points out that the countryside has traditionally been an image of the past, the city is the image of the future, and the industrialisation of society was 'a kind of fall, the true cause and origin of our social suffering and disorder'. But it is early days yet, of course; our urban time is nothing compared with our centuries 'close to nature'. Might we yet overcome the human failing of never knowing where we stand, or what we should expect of the planet? Or are town and industry so inherently unpalatable to our spirit that we shall never become comfortable with them?

In order to come to terms with our perceived ecological predicament, Schama (1995) wonders if 'a new set of myths are what the doctor should order as a cure for our ills'. For much of what we think we have lost only ever existed in the mind and should therefore be recoverable if we review our aspirations.

> the cultural habits of humanity have always made room for the sacredness of nature . . . So that to take the many and several ills of the environment seriously does not, I think, require that we trade in our cultural legacy or its posterity. It asks instead that we simply see it for what it has truly been: not the repudiation, but the veneration, of nature.

Schama reviews our attitudes to different types of landscape, showing how greatly they vary with both time and place. For example, he puts the case that forests are associated in France with orderliness, in Germany with militarism, in Poland with the struggle for national independence, in America with man's proximity to God, in England with personal freedom. Yet, while forests may mean that to some people, others feel hemmed in by them, clothed by their benevolence or malice, depending upon the light and mood. Freedom may more easily be felt upon the open moor. 'Landscapes are culture before they are nature; constructs of the imagination projected onto wood and water and rock,' says Schama. The imagination 'where the wildest of myths have insinuated themselves into the lie of our land'.

It may very well be a lie, but it is upon the inspiration of the natural world that the wish to conserve has been based. There is absolutely no substitute for experiencing nature at first hand. For those who have not been recipients of this basic human right, the enthusiasm of others can be hard to comprehend. Countryside by proxy, through magazines or television, cannot put it right. Academic expertise can miss the magic, either because too much work is laboratory-oriented or because the specialisation runs too deep. We shall see (p. 33) how nature study was to become a social pastime, but the field naturalists soon came to appreciate its limitations.

> Faunists, as you observe, are too apt to acquiesce in bare descriptions, and a few synonyms: the reason is plain; because all that may be done at home in a man's study, but the investigation of the life and conservation of animals is a concern of much more trouble and difficulty,

and is not to be attained but by the active and inquisitive, and by those
that reside much in the country.

(Gilbert White, in a letter to Barrington, 1771)

Philosophers more grave than wise
Hunt science down in Butterflies;
Or fondly poring on a Spider
Stretch human contemplation wider;
Fossiles give joy to Galen's soul,
He digs for knowledge, like a Mole;
In shells so learn'd, that all agree
No fish that swims knows more than he!

(John Gay, 'To a Lady on her Passion for Old China', 1725)

Industrialisation removed people from nature for the first time in the
history of the human race. Furthermore, it dehumanised them by depriving
them of individualism. Work was mechanically oriented towards the
requirements of machines; living conditions settled at the lowest common
factor of large-scale urban existence. Knowledge was divorced from values,
fact from feeling, head from heart. Inevitably, wild nature began to assume
a fantastic attraction. 'Thou hast pined and hungered after Nature, many
a year, In the great City pent', wrote Coleridge. It is this pining which,
today, drives urban dwellers to the countryside every weekend, in spite
of the traffic jams. It is this pining which sells 'country-fresh' produce,
glossy magazines and television programmes. It matters not to urban
dwellers that what they seek has never existed; the spirit feels a need for
it and the search can answer much of that need. We have expended a
great deal of energy in the quest. In the nineteenth century, it gave rise
to the garden city movement and to the great British holiday. More
recently, it has led to city suburbs and dormitory towns, to satellite settlements and second homes. Its keenest followers are also the staunchest
supporters of the conservation movement, fighting off threats to their own
back yards and seeking to retrieve a countryside that has never been.

> Thousands of tired, nerve-shaken, over-civilised people are beginning
> to find out that going to the mountains is going home; that wildness is
> a necessity and that mountain parks and reservations are useful not
> only as fountains of timber and irrigating rivers but as fountains of life.
> Awakening from the stupefying effects of the vice of over-industry and
> the deadly apathy of luxury they are trying as best they can to mix
> and enrich their own little ongoings with those of Nature, and to get
> rid of rust and disease.

We were still in the nineteenth century (1898) when John Muir, the pioneer of American National Parks, wrote these words which ring so true to
this day in all developed countries. For the towns were not beautiful places
in which to live. Functional workplaces, they were the hub of the world's

first industrial nation and harbingers of wealth and influence. The British people had much to be thankful for and we cannot look back in technological judgement and belittle the industrial towns in any way. But they were not beautiful places – and the human spirit needs beautiful places. So there were protests about the squalor, the suffocating smogs and the polluted watercourses; protests that all the salmon had disappeared from the Thames between 1820 and 1830; protests against the effluent filth of the collieries and lead mines. In response, there was an 1863 Alkali Act and a Public Health Act of 1875. Most of our urban parks were developed at this time to appease the demands for breathing space. Even the dead were allowed to rest in peace and solitude in the new cemeteries; previously their bones had been exhumed for grinding into fertiliser.

But the headlong rush for industrialisation could not afford to falter in the face of protests over pollution, so in addition to the protests, there was a polishing of 'the myth of a happier past' (Williams 1973); the 'grass is greener' syndrome. Not out of concern for natural history, of course, but for human social and spiritual reasons. Blind eyes were turned to the mundane realities of rural life, just as they are today. Constable only painted what the mind's eye wanted to see. The back-breaking graft was not shown to complement the hell of factory work; the mean cottages were not put upon the same low level as the slums. The countryside, from which people had so recently escaped, took on an aura of idealism; of freshness and freedom, of health and happiness. It had little to do with reality, still less to do with nature conservation. 'Yet, despite all its falsities, the growing rural sentiment reflected an authentic longing which would steadily increase, both in volume and intensity, with the spread of cities and the growth of industry' (Thomas 1983). It is an aura which the countryside wears still and upon which much of the modern conservation movement has been built.

In the light of these new attitudes, the wisdom of taming the natural world was called into question. For hundreds of years, our dominion over nature had been an unquestionable aim. Now, just as industrialism seemed set to present unbounded possibilities of power, moral considerations moved the goalposts. With the publication in 1859 of *On The Origin of Species*, and in 1871 of *The Descent of Man*, Charles Darwin reduced human beings to the level of all creatures. Suddenly, the intimidation of nature took on more personal overtones. Living largely divorced from it, the new urbanites began to respect its value. Together with the growing concern for the welfare of animals, this hardened into a firm base for preserving the countryside.

Railway mania in the 1830s and 1840s saw the construction of almost 13,000 km of track. Towards the end of the century, the bicycle became a viable means of personal transport. These innovations provided the urban population with ready access to the countryside and made its preservation a more personal and meaningful affair. The personal aspect was reinforced by upper-class assertions that the urbanites would fail to treat the countryside correctly. There was opposition to public access, and the first of many local footpath societies was formed in 1826. It was the sapling stage of another chestnut of countryside conflict that is with us to this day.

What these urbanites found in their eager return to the countryside was not entirely what they had expected. Much of this was due to the myths they had made for themselves. But change was occurring and agriculture, like the towns, was moving away from nature. Attitudes altered accordingly. Thomas (1983) quotes William Gilpin, at the beginning of the nineteenth century, who states: 'Wherever man appears with his tools, deformity follows his steps. His spade and his plough, his hedge and his furrow, make shocking encroachments on the simplicity and elegance of landscape' and goes on to submit that England would 'be more beautiful in a state of nature than in a state of cultivation. . . . The regularity of cornfields disgusts and the colour of corn, especially at harvest, is out of tone with everything else'. Replace the word 'regularity' with 'size' and the word 'corn' with 'oil-seed rape' and you have the epitome of today's argument against intensive agribusiness. We all wish to eat home-grown food but the sight of it being grown on any scale is displeasing to many eyes. This nineteenth-century attitude settled on the farmer's ground and has lain there ever since. Only the details alter a little. The hedgerows then were seen as monotonous and unnatural, while today we fight tooth and nail to preserve them. In fact, John Stuart Mill, four years after the last Enclosure Act, bemoaned the widespread uprooting of hedgerows in the name of agriculture. He saw their removal as part of farming's master plan to tame wild nature and deprive the human spirit:

> Solitude in the presence of natural beauty and grandeur, is the cradle of thoughts and aspirations which are not only good for the individual, but which society could ill do without. Nor is there much satisfaction in contemplating the world with nothing left to the spontaneous activity of nature; with every foot of land brought into cultivation, which is capable of growing food for human beings; every flowery waste or natural pasture ploughed up, all quadrupeds or birds which are not domesticated for man's use exterminated as his rivals for food, every hedgerow or superfluous tree rooted out, and scarcely a place left where a wild shrub or flower could grow without being eradicated as a weed in the name of improved agriculture.

A late twentieth century sentiment, one might think, but in fact he wrote it in his *Principles of Political Economy* of 1848. It was a far cry from the scornful damning of wilderness, heard only two generations earlier.

Enclosure had changed the face of the countryside. The process was nothing new to England; 'fields demarcated by hedges, walls, ditches, or banks have been typical of the British landscape since civilization began'

(Rackham 1986). Many of them were the remains of primeval woodland cleared during prehistoric times; others dated from the Romans or Anglo-Saxons. Following further enclosure during the thirteenth and fourteenth centuries, 50 per cent of England had been enclosed by the start of the sixteenth, including most of the western counties and the south-east. But in the hundred years before Mill was writing, 325,000 km of new hedgerow was planted. Over 1.2 million ha of open farmland and 800,000 ha of heath, common and other marginal land were altered dramatically by over 4,000 Enclosure Acts. The British landscape became one of pocket-sized fields trimmed with ineluctable hedgerows. In their heyday, 1 million km of hedgerow covered 1 per cent of Britain. Their impact on our wildlife was enormous. Some types suffered but, as an extension of the woodland cover which had been disappearing for almost 5,000 years, hedgerows proved very advantageous to most native species. It has been estimated that ten million birds still breed in our present much reduced network of hedgerows. Much of our flora, too, has taken shelter under the hedges from the advances of modern agriculture.

While the new hedgerows benefited much of our wildlife, nature was not in the minds of the legislators or the farmworkers who made them. It was a time of agricultural intensification, and the nineteenth century saw some advances of its own. In 1870, there were 5 million ha of arable land in England, a figure which was to be exceeded only once in the next one hundred years. Stamp (1969) sees it as 'the period of "high farming" in Britain, which . . . may be taken as a starting point for the consideration of the modern interaction between farming and wildlife'. Any increased use of the countryside for human ends was bound to impose new conditions upon its wildlife. 'Agriculture, after all, consists of distorting the ecosystem in favour of man' (Kenneth Boulding, quoted in Chisholm 1972). Any animal or plant that could not adapt would suffer and one casualty of the time was the great bustard. This large, turkey-like running bird was first recorded in the fifteenth century and had roamed in enormous flocks across the open spaces of Salisbury Plain, East Anglia and Yorkshire. Field enclosure deprived it of its preferred habitat; the introduction of the horse-hoe left the eggs susceptible to trampling or collection for food; and the change from rye to wheat crops upset the bird's nesting habits. These purely agricultural factors, together with improvements in sporting firearms – the bird was turkey-like to taste as well – spelt its rapid demise. It disappeared from Suffolk (1812), Wiltshire (1820), Yorkshire (1830) and finally Norfolk (1838).

Other changes in farming techniques affected other types of wildlife and other habitats. We begin to realise that threats from agriculture are nothing new, and neither are the conservationist protests that they provoke. After a spell of increased planting for timber, we find more trees being felled for agriculture between 1840 and 1880. William Cobbett was bemoaning the tearing up of 'wasteland'. In the nineteenth century,

Plate 2.2 Enclosure was unpopular for social reasons and because it changed the landscape, but it did give us some of our valuable hedgerows: medieval field patterns on Culm Davy, Devon. Photograph by Peter Hamilton. Courtesy of Countryside Commission

farmland was being fertilised with chemicals, albeit organic ones. In the nineteenth century, government grants were available for drainage work.

> 1851 was a black year: the year we lost the unique Whittlesey Mere (Huntingdonshire) and Hainault Forest (Essex) for merely a few more farms, some of which proved to be temporary. Protests at the destruction of Hainault were the foundation of the modern conservation movement, but the short-sighted Victorians acquiesced in the destruction of Whittlesey Mere. Since 1945 there have been some twenty-five years each as bad as 1851.
>
> (Rackham 1986)

The deleterious effects of drainage work were nowhere more obvious than in the fenlands of England. In the first fifty years of the 1800s, 275,000 ha of fenland in East Anglia were drained and it was this period that accounted for the greatest losses in both wetland habitat and its characteristic species. By the end of the century, the bittern, spoonbill, greylag goose, marsh harrier, ruff, black-tailed godwit, avocet, black tern and Savi's warbler had all been lost as breeding birds. Shooting for sport, collection of both birds and eggs, and increased disturbance all played a part in these losses but the drainage of wetlands was the chief cause.

Intensive conservation has been needed to bring back some of these birds as breeding species: the bittern (1911), marsh harrier (1926), avocet (1947), Savi's warbler (1960), black tern (1966), ruff and black-tailed godwit (1971).

Conservation measures on behalf of the large copper butterfly have not been so successful. First identified as a species in 1795 and very common in the fens of Huntingdonshire at the start of the nineteenth century, it was completely exterminated there by drainage work in the 1840s. Collectors may have picked off the final handful but they cannot carry the blame for its colossal decline. Efforts in the 1920s and since the 1970s to introduce the Dutch race to certain areas, notably Woodwalton Fen, have been unconvincing in their results in spite of extensive management work to control the scrub of sallow, birch and alder in favour of reed and sedge (p. 106). Most of the nineteenth century effects upon our countryside have proved irreversible.

Forestry, too, had outgrown the whims of the landed gentry and was being put on a large-scale, organised footing. Almost inevitably, attitudes railed against the monoculture of larches in symmetrical, sharp-edged plantations. Wordsworth referred in disdain to the 'vegetable manufactory' of the Lake District. The oldest forestry society in the English-speaking world dates from this time: the Royal Scottish Forestry Group was set up in 1854 with the stated aim (apparently ignorant of the pun) – 'the advancement of forestry in all its numerous branches'.

SPORTSMAN AND COLLECTOR

Strange as it may seem today, nobody bemoaned the passing of the avocet or the large copper. In reality, lack of serious recording meant that such species were never missed. Mere casualties of the changing times, of little general use or interest, they had no champions to support their cause.

The demise of other species was more closely followed. The continuing popularity of the 'country sports' was another aspect of nineteenth-century society which was of great consequence to British wildlife. Improvements in the mechanisms of popular firearms were rapid and impressive, and breech-loading guns soon ousted the muzzle-loaders. This meant a saving in time and effort when loading, and in powder which no longer failed through being wet. To the quarry, it meant more blasts per minute. The new guns also enabled the user to fire upwards at birds overhead, a practice fraught with danger when using a muzzle loader. 'It was a revolution that was to have a profound effect upon British wild life' (Vesey-Fitzgerald 1969). As well as changes in the tools of the trade, the rules of the game were altered in the latter part of the 1800s. Rough shooting with dogs working the ground before the guns lost favour to the use of beaters driving the quarry towards them. Again, some time and much effort were saved. To the quarry, this meant more hits per hundred blasts.

Furthermore the new principles encouraged the growth of the sporting estate with its massed guns and its relentless gamekeepers.

But before we turn to this very British institution, let us not underestimate the power of the single gun: the lone 'sportsman–naturalist–collector' who epitomises the nineteenth century.

> What a big pile it would make if all the blackgame I shot between 1855 and 1900 were gathered in one heap. Now alas! there are none, and why, who can tell.
>
> (Osgood MacKenzie, *One Hundred Years in the Highlands*, 1952; quoted in Perry 1978)

> There are but very few [ospreys] in Britain at any time, their principal headquarters seeming to be in America; and though living in tolerable peace in the Highlands, they do not appear to increase nor to breed in any localities excepting where they find a situation similar to what I have already described [an islet in a lonely loch]. As they in no way interfere with the sportsman or others, it is a great pity that they should ever be destroyed.
>
> (Charles St John, *A Tour in Sutherlandshire*; quoted in Vesey-Fitzgerald 1969)

St John was writing after having whiled away the summer of 1848 shooting Scottish ospreys. He killed two adults, shot at many more, and added four young birds and three eggs to his collection. Vesey-Fitzgerald says that, as men of their time and not of ours, they should not be condemned. But their ignorance of reality seems too blatant and wilful. No, there was nothing amiss in shooting ospreys in the nineteenth century, but then to ask such rhetorical questions was severely supercilious. It is an unfortunate attitude which has lingered on in certain quarters to this day.

Punt-gunning was a lesser-known practice but its local effects on some populations was staggering. One practitioner in Norfolk claimed 603 knot, nine redshanks and five dunlins with just two shots!

There had been a number of Acts over the years – 1671, 1723, 1770, 1803, 1816 – which had attempted to counter the traditional 'country sport' of poaching. The frequency of these Acts, together with the fact that one in every seven convicted criminals in 1827 was a poacher, tends to betray their ineffectiveness. The Game Act of 1831 gave legal recognition to the profession of gamekeeper which had its advent with the enclosures of the late 1700s. Keepers became respected persons in the community and were given powers to arrest. Their duty was to protect from all and everything (except the sportsman), hares, heath or moor game, black game, pheasants, partridges and bustards. It seems that no effort was spared, particularly against our birds of prey. Their decline was spectacular in all respects; the methods used, the speed, the geographical areas affected and the sheer numbers.

Nineteenth-century persecution drove the golden eagle out of northern England and Wales. The bird bred in Ireland until 1912. The white-tailed sea eagle had been widespread in eighteenth-century Scotland and Ireland with a few pairs breeding in England. By 1916, it was extinct throughout Britain and Ireland. The same year witnessed the last breeding pair of ospreys. Buzzards, having been persecuted since the seventeenth century, numbered about 12,000 pairs at the start of the nineteenth. They eventually died out in Ireland during the 1880s and by 1915 only a remnant population survived in the north and west of Britain. To this day, persecution prevents their recolonisation of eastern parts. 1870 saw the last nests of the red kite in England and, by the turn of the century, numbers in Wales may have dropped to as few as five individuals.

Perry (1978) quotes spine-chilling figures for birds of prey killed on one estate, Glen Garry in the Grampians, in just three years, 1837–40: 1,796 birds, namely fifteen golden eagles, twenty-seven sea eagles, eighteen ospreys, sixty-three goshawks, 275 kites, seventy-two hen harriers, five Montagu's harriers, ninety-eight peregrines, eleven hobbies, six gyrfalcons, seven orange-legged falcons (sparrowhawk), seventy-eight merlins, 462 kestrels, 285 buzzards, 371 rough-legged buzzards and three honey-buzzards. Together with 109 owls and 475 ravens. Even allowing for a gamekeeper's exaggeration, such numbers are fabulous by modern standards.

And so the story goes on. Very common at the start of the nineteenth century, marsh harriers were extinct in Britain by 1900 and in Ireland by 1917. The end of the 1800s left a few hen harriers in Orkney, the Outer Hebrides and in Ireland. Montagu's harriers had never been common and, before the end of the nineteenth century, they were down to three or four pairs, these in Norfolk. The goshawk had gone from Scotland by the 1880s and the last breeding birds in England (Yorkshire) were shot in 1893. Numbers of sparrowhawks and all our falcons were severely reduced although these were the more successful of the raptors in riding the storm.

Birds were not the only predators of game species and their eggs. The pine marten had been driven from most of lowland England by 1850. Running alongside persecution by gamekeepers, loss of their pine forest habitat reduced numbers still further and forced them to move to the treeless moors of Scotland and Wales. But these moors were home to the grouse and persecution followed them there. The wild cat was driven from the high places of England and Wales by the end of the century and survived only as a remnant population in northern Scotland. Polecats were hunted by dogs until mid-century. Thereafter, continued persecution brought their numbers to a record low by 1900, with the main concentrations hanging on in Wales. Perry (1978) does not understate the fact when he says that, from the mid-1800s until the Second World War, 'the survival or otherwise of Britain's fauna was determined predominantly by the landed proprietors and their gamekeepers. Britain became the most intensively keepered country in the world'.

While the effects were greatest in lowland areas, field sports were also helping to mould the highlands. The Scottish countryside particularly was in a state of flux. Although the 1745 Jacobite rebellion had failed, it resulted in an opening up of the country and much shifting of population. The English began to arrive in numbers to participate in the grouse-shooting. The consequent control of predators and regular burning of heather made the moors even less natural. In other areas, sheep were introduced and quickly altered the flora and soil characteristics. Vast tracts of the Scottish countryside today are the result of these two activities.

Having touched upon the effects of the gamekeeper with his gun and his traps, it is perhaps surprising to see Stamp (1969) asserting that 'modern conservation really begins with the famous Game Act of 1831'. Conservation of a handful of species in the interests of sport is not the modern view of the science. But it did focus attention on the problems of preserving species and maintaining habitats. It did highlight the inter-relation between various types of animal and plant. And it did introduce nature conservation to the more ancient arena of country sport. The inter-action between the two has been intimate and is one of the broadest issues in conservation today. The gamekeeper became possibly the most knowl-edgeable of countrymen but was paid to use his knowledge against the predators. He was the main character in the nineteenth-century British countryside. None of the creatures mentioned so far has regained its former status, even though keepering has declined since the 1930s. And the conservationist continues to smart under the potency of the Game Act and its successors:

> game laws continued to be harsher than common sense would suggest right down to the present day. . . . laws that protect partridges and pheas-ants more severely than owls and songbirds seem out of place in the climate of late 20th century Europe.

> (Fitter and Scott 1978)

What then were the lesser evils affecting our wildlife in the 1800s? Thanks to the growth of museums and private collections, specimens were in sufficient demand to make collecting them a profitable business. While some scientific study was carried out, most private collections were a social phenomenon brought on, perhaps, by the Briton's new-found isolation from nature. The race to keep up with nineteenth-century neighbours revolved, in some quarters at least, around cabinets of eggs and glass cases of ferns, stuffed animals and birds. In 1860, a pair of honey buzzard eggs would fetch £5. One June day in 1870, the illustrious 'ornithologist' Henry Seebohm took 456 eggs home with him from nests on the Farne Islands.

In 1884, Bempton Cliffs, now an RSPB reserve, lost 130,000 guillemot eggs to collectors. Perhaps the collector never caused a bird to become rare, but rarity was his best business, with the vested interest of high prices to keep them rare.

In the case of the great auk, he overstepped the mark. Always harvested as food, its eggs and skins then became desirable to collectors and the last British specimen was shot in Scotland in 1812. The bird was extinct worldwide by 1844. But collectors maintained their interest and prices soared. In 1882, Lord Lilford bought a bird and egg for £300 from a fellow collector, and twelve years later, an egg alone fetched 300 guineas at auction in London.

Beachcombing also became a lucrative pastime. Amateur collectors took vast numbers of seaside creatures and plants. Sea-anemones and crustacea, seaweeds and ferns suffered in particular. But the real damage was done when commercial interests inevitably invaded the market. Fortunately for our coasts, the collecting craze, on that scale at least, eventually waned but it took many decades for the worst-affected species to recover.

Collecting live birds was lucrative too, especially as the new-found interest in God's creatures encouraged the keeping of songbirds. In the 1840s, 22,000 wheatears were caught annually as their autumn migration took them across the downs of Sussex. Perry shows how they were trapped in a loop of horsehair strung between two sticks which supported a previously cut turf. As the bird moved forward, the turf collapsed and the quarry was taken. Such effort on the part of the trapper can only have been made with an expectation of high reward. In the 1860s, at Worthing in Sussex, 132,000 goldfinches could be taken annually. By 1898, there were thirty-three pairs of bearded reedlings in Norfolk and a handful on the south coast. Earlier in the century, they had been widespread throughout east and south-east England. At the start of the century, the woodlark bred in most of England and Wales and in parts of Ireland. By the 1850s, it had disappeared from north-west England and Ireland, never to return again as a regular. The catalogue of destruction is a hefty one and can only be touched upon here. Suffice it to say that many of our songbirds faced catastrophic decline from which they have never recovered.

LAW AND SOCIETY

Fortunately, the changing attitudes that heralded this new-found interest in wildlife had more positive outlets as well. In the eighteenth century, the middle classes who had the leisure, the wherewithal and the inclination, began to form themselves into natural history societies. Albeit that they were typically a type of social club, it was thanks to them that interest in natural history developed a broad and secure foundation. Otherwise, it might have become a purely scientific discipline, as inaccessible to the amateur as many other branches of science have become.

The earliest society of this type in fact pre-empted the century by a decade. Formed in 1689, the Temple Coffee House Botanic Club was a manifestation of the social aspect of collecting plants. This interest arose from their use in the medicines of the day and the need to supply the botanic gardens which inevitably sprang up as a result. In 1721, Dillenius's Botanical Society began its five year existence. The Aurelian Society for insect study lasted only three years of the 1740s. A Society for Promoting Natural History was active for ten years from 1782. But not all were so short-lived. In the year that saw the publication of Gilbert White's classic *Natural History of Selborne*, the Linnaean Society was established on 26 February 1788. It thrives to this day. Its Scottish counterpart, the Wernerian Society of Edinburgh, followed in 1808.

The short life of most of these groups matters not. What is important is that they were formed at all; formed out of new and genuine interest in the natural world. It is the first evidence of such an interest standing alone: the study – though not the conservation – of nature as an end in itself. No genuine interest would be knocked back by a few false starts. Thus the failure of the Aurelian Society was followed by attempts to revitalise it in 1762 and again in 1801. Insect study continued regardless. Wherever a need was identified, a society was formed to fulfil it, and that process continues to this day. Societies have been the organ by which natural historians and, later, conservationists have come together to pool their expertise and focus their aims, to benefit from their communal strength and sometimes to shut the doors on other interests. Societies have epitomised the British scene and, in the world of natural history, their roots are bedded in the eighteenth century.

But it was only during the last thirty years of the nineteenth century that the first stirrings of a nature conservation movement could be discerned. In the face of unique threats to nature – habitat loss, shooting, collecting – something had to be done in its defence. General attitudes were still such that the law was seen as the best hope for some immediate relief. A handful of active conservationists were remarkably successful in pushing through a number of new laws. This success was less to do with enlightened attitudes than with the influential positions of many of those who championed the cause. The year 1888 saw the first local by-laws for the protection of plants on a county basis, but birds enjoyed the favour of just about all other early efforts. Birds, because they were obvious and colourful; birds, because they attracted attention, though not always the best sort of attention. There was a great fashion amongst the ladies of the time for wearing feathers in their headgear. The home supply, though great, could not satisfy demands and international trade was fast and lucrative. Egrets and herons, gulls and terns, pheasants, hummingbirds and birds of paradise were all at risk. In the 1790s, fishermen in the small village of Clovelly in Devon were collecting 9,000 kittiwake wings during an August fortnight. The true cost of the millinery trade at this time will remain forever unfathomed.

Owing to the fact that the Royal Society for the Protection of Birds (RSPB), as it eventually became, grew out of the fight against the plumage trade, it is often forgotten that there were many other threats to our birds at this time. Shooting for collections was taking its toll, but shooting for 'sport', in the loosest sense of the word, was also rife. Parties of 'sportsmen' would take out a pleasure steamer along the coast or down a river from which to blast at anything that moved. The Yorkshire Naturalists' Union started the legal ball rolling by making calls for the protection of the seabird colonies on Bempton Cliffs and Flamborough Head. And in June 1869, the Sea Birds Preservation Act introduced a close season from 1 April to 1 August on thirty-three species.

So began a long series of bird protection Acts which has left us today with the most comprehensive protection laws in the world for one class of animal.

> Bird protectionists ... may claim to have been the earliest and, at least until very recently, the most energetic and successful arm of the world conservation movement.
>
> (Nicholson 1970)

The 1876 Wild Birds Protection Act extended the close season and the number of species covered. Four years later, both Acts were replaced by another Wild Birds Protection Act which covered all species between 1 March and 1 August. This Act survived without amendment for fourteen years, although a special Act of 1888 did afford protection to the St Kilda wren, a subspecies of that tiny bird confined to that tiny island. The 1894 Act, when it came, changed much while keeping the same title. It enabled county councils to introduce protection of certain species. Thus the goldfinch came to be protected in Somerset, for instance. It was also the first measure that extended to eggs, being drawn up at a time when egg collecting was rampant. The Act invited councils to identify 'reserves' in which all species and their eggs would receive protection. Further Wild Bird Protection Acts of 1896, 1902 and 1904 consolidated earlier measures and prove beyond doubt that the conservationists were piling on the pressure.

However, it is never plain sailing in legal waters, and the craft of early bird protection was buffeted by derision and riddled with loopholes. The number of cases brought against offenders was disappointingly low. Potential fines were easily outweighed by the profits to be made. Fines under the 1876 Act were so low that they were reviewed before the year was out, yet they still remained derisory for the time and the crime. It remained lawful to possess birds taken outside the close season or outside the country, so once the quarry had been reduced into the possession of

Plate 2.3   Only coming onshore to breed, many of our seabirds ought to have been safe from the hands of man, but shooting for 'sport' was superseded by pollution of the oceans and now by depletion of natural foods by overfishing: puffins on Shetland. Courtesy of Countryside Commission

the catcher, it was impossible to prove an offence and trading could continue with impunity. Until 1894, eggs were unprotected at a time when vast numbers were taken annually. Inertia was strong, so popular was the pastime, and many councils failed to introduce protective measures of any sort. Likewise, very few reserves were identified. Those that were – the Farne Islands and Skokholm, Dungeness and the Norfolk Broads, for instance – immediately attracted the greater attentions of the collectors. Later moves to enable courts to confiscate birds, eggs and equipment proved no real deterrent.

The problem, quite simply, was one of attitudes. Bird catching, egg collecting, open trading in both; these things were a way of life as common as poaching or the newspaper vendor on the street corner. Overnight, they became legally wrong, but moral principles changed more slowly. The introduction of the laws was a surprising success; their immediate effect in real terms was unsurprisingly limited.

The conservationists faced a long haul. They were a tiny group. The scientists were taking a new path into the field of physiology and structure. Individual species hardly mattered to them; field study was irrelevant. The aesthetic aspect of nature was drowned under the weight of scientific fact; its spiritual value failed to show up under the microscope. When the Education Act of 1870 supported the inclusion of Science on school timetables, it was this sort of impersonal science that was introduced. 'Nature Study' came to mean the dissection of dogfish in the classroom; biology considered configurations of cells. Many potential conservationists shuddered and turned away for life. And the situation has remained that way to date. 'Well before the end of the nineteenth century, the study of the natural environment had ceased to be purely academic preserve' (Bonham-Carter 1971). This was because the conservationists took their own line in nature study. Like any minority group, they looked to each other for mutual support and this manifested itself in the formation of the societies. Typically, whether local or national, these early societies were field clubs, concerned with a first-hand study of nature. They also aimed to mobilise public opinion on behalf of our wildlife. It is difficult now to envisage what a small embattled core of diehards it was that promoted these groups. But from them arose many of our most influential conservation bodies today.

1824 saw the setting up of the Society for the Prevention of Cruelty to Animals, which became 'Royal' (RSPCA) sixteen years later. Its formation highlighted the growing concern for animal welfare and, although involved primarily with domesticated creatures, it was active in some early crusades of wider conservation significance, notably its support for the 1869 Sea Birds Preservation Act. In the same vein, the Zoological Society of London (1826) was concerned with both captive and wild animals.

The Entomological Society of London was formed in May 1833 from the ashes of the eighteenth-century groups (p. 34). Its object was 'the improvement and diffusion of entomological science' and its Royal Charter was granted in 1885. In its centenary year, it became the present Royal Entomological Society (RES).

The Botanical Society of London was added to this family of associations in 1836. Much exciting survey work was being carried out on our flora at this time. Individuals produced the earliest atlases of plant distribution: William Brand in 1838 and H.C. Watson in 1843. The latter went on to develop his extensive *Topographical Botany* of 1873–4. During the 1830s, efforts were made to combine the work of individual local surveys into a large-scale atlas. A similar idea to trace bird migration through local survey failed to attract interest. 'British zoology, unlike British botany, had still to wake up to the possibilities of collective research' (Allen 1976). However, after the Society became the Botanical Exchange Club in 1858, it sunk into oblivion. Revitalised this century, it eventually became the present Botanical Society of the British Isles (BSBI) (p. 88).

A number of books were now available and these played a great part in furthering interest in natural history. Gilbert White's *The Natural History of Selborne* (1789) was a century ahead of its time and remains a classic to this day for its astute observations of wildlife. Famous naturalists, such as Charles Waterton, were writing profusely. In 1850, Rev. F.O. Morris began his partwork, *A History of British Birds*, followed two years later by *A Natural History of Nests and Eggs of British Birds* and *A History of Butterflies*. By the end of the century, other great naturalists were in print: Lord Lilford, W. Warde Fowler, W.H. Hudson.

In 1858, the British Ornithologists' Union (BOU) was formed and supplemented, in 1892, by the British Ornithologists' Club (BOC). Although largely scientific and theoretical in its approach to birds, the relevance here is that the BOU is now the oldest ornithological society in existence.

The Conchological Society of 1874 originated from that section of collectors who did have a genuine interest in seashells. Likewise the Marine Biological Association of the UK (1884) catered for those with more than a social interest in the topic. The 1890s saw new societies covering a range of interests, such as fungi, mosses and ferns.

By this time, many parts of the country had their own local societies, such as the London Natural History Society set up in 1858.

The first nationwide association to be concerned with all forms of wildlife was the Selborne Society for the Protection of Birds, Plants and Pleasant Places. Formed in 1885, its aims were to:

- preserve from unnecessary destruction such wild birds, animals, and plants as are harmless, beautiful or rare;
- discourage the wearing and use for ornament of birds and their plumage, except when the birds are killed for food or reared for their plumage;
- protect places and objects of interest or natural beauty from ill-treatment or destruction;
- promote the study of natural history.

The second aim accounted for much early effort, particularly between 1886 – when the Society coalesced with the year-old Plumage League – and 1889 when the Fur and Feather Group was formed and took the helm in the campaign. The Selborne Society lives on to this day.

The other burning issue besides nature study and protection was, as we have seen, the question of public access. The Commons Preservation Society (CPS) was launched in 1865 by John Stuart Mill, Octavia Hill, Sir Robert Hunter and others. It arose out of concern over the restricted access to much open land around our cities and the continuing development or enclosure of common land. Commons, which exist only in England and Wales, are owned either privately or by public body and may not be enclosed. But there is no automatic right of public access. Of the 600,000 ha that exists today, only 25 per cent is open to the public. The battle has been a long one and goes on today for much the same reasons as a hundred years ago: the increased mobility of the urban population who wish to spend their spare time close to nature. CPS concentrated its early efforts, not surprisingly, on the green spaces around London and succeeded in keeping open valuable lungs such as Epping Forest, Blackheath, Hampstead Heath, Wandsworth and Wimbledon Common. Present Londoners have much to thank it for.

In the last year of the century, it joined forces with the National Footpaths Preservation Society to become the Commons, Open Spaces and Footpaths Preservation Society. After many years of labouring under the weight of that name, it has now become the Open Spaces Society (OSS), a name much more redolent of the freedom that it champions.

The struggle for access is inextricably part of the nature conservation movement. Too much public access may be detrimental to animals and plants; it may spoil the vegetation cover and detract from the landscape value. But public use of an area is a powerful argument against more damaging developments and has saved many an area of high wildlife value. Public access encourages education and increases the number of caring people with a positive will to conserve nature. Adams (1986) is in no doubt as to its importance. Talking of the early work of the CPS around London, he says:

> This work was the first on any scale which involved the protection of particular places against development. From it grew the idea of conserving nature by protecting particular areas and habitats in natural reserves or on private land. This has consequently become the dominant feature of British conservation.

Although commons do not exist in Scotland, the struggle for access most certainly does. In 1884, James Bryce MP proposed free access to all uncultivated mountain and moorland in Scotland. His Access to Mountains (Scotland) Bill failed, on all twelve occasions that he promoted it, but it set the ball rolling for a movement which is still battling against powerful vested interests. In 1908, 1926, 1927, 1928, and again in 1931 and 1937, similar bills covering the moorlands of England and Wales were tried and failed, the first four of them championed by Charles Trevelyan. The move into the mountains was going to be an uphill slog in more ways than one.

So where did conservation stand by the end of the 1880s? In the face of fundamental changes to the rural landscape, but largely due to the type of urban surroundings in which an increasing number of people were living, attitudes were altering on mankind's treatment of the natural world. In the face of devastating inroads being made into the numbers and variety of wildlife, notably birds, some protective laws were in force. To capitalise upon the growing interest, new societies were springing up overnight. But the attitudes did not change the way in which the landscape was managed. The laws were rarely used in practice, and found to be ineffectual when they were. The societies were not always a success and, whilst their proliferation was on one hand vitally important, it was beginning

Plate 2.4 The uphill struggle for freedom to roam the open spaces has now entered its second century: above Glen Einig, Cairngorm. Photograph by Jo Porter. Courtesy of Countryside Commission

to turn cumbersome on the other. Toes were being trodden upon in the headlong rush of enthusiasm to pursue broadly common aims: namely, the encouragement of nature study, the preservation of beautiful places and, most specifically, the battle on behalf of birds against the millinery trade. It was fortunate that jostling gave way to some joining of forces. The result, as we shall see, was a great success and our present conservation societies might do well to heed the lesson. In the quarter-century that pivots about the year 1900, three of our most influential and successful societies were formed. They all became 'world-leaders' in their first hundred years, spearheading the British campaign to make nature conservation meaningful and effective.

3   1890–1940: Societies and suburbia

LEADING LIGHTS

The Royal Society for the Protection of Birds, together with its junior section, the Young Ornithologists' Club, has a membership in excess of 1 million. It is the largest voluntary nature conservation body in Britain. Christened the Society for the Protection of Birds in 1891, it grew directly out of the Fur and Feather Group set up in Manchester on 17 February 1889 by solicitor's wife, Mrs Emily Williamson. The merging with other anti-millinery groups brought about the change of name but the movement continued an almost entirely feminine one in its early years. It was based wholly on the emotive nature of the feather trade which killed birds cruelly and left their defenceless young to starve unattended in the nest. A constitution was formalised, ruling that:

- members shall discourage the wanton destruction of birds and interest themselves in their protection;
- lady members shall refrain from wearing the feathers of any bird not killed for the purposes of food, the ostrich excepted.

Right from the start the society attracted tremendous interest, not least of all through the personal persuasion of its often influencial champions. Moreover, the general attraction of bird life was on the increase. The winter of 1890–1 was particularly hard and saw nationwide appeals exhorting people to feed the birds. When the snows had thawed, it became clear that so too had the hearts of urban Britain. The feeding continued, bird tables flourished and nestboxes multiplied. An 1897 booklet, *Wild Bird Protection and Nesting Boxes* ironically suggested an empty gun-cartridge box as an ideal nesting site for tits. Without doubt, the increasing bewitchment of the British public by the birds played a part in the good fortunes of the RSPB. By the end of the century, membership stood at over 25,000, which by then included many men, and a paid secretary was employed.

The royal charter was granted in 1904 and outlined the two main aims of the society:

- to encourage the better conservation and protection of wild birds, more particularly of rare or interesting species, by developing public interest in their place in nature, as well as in their beauty of plumage and note;
- to discourage the wanton destruction of birds and the wearing of feathers of any bird not killed for the purposes of food, other than the ostrich, but to take no part in the question of killing game birds, and legitimate sport of that character.

The first 'watchers' were appointed in 1901 to protect breeding pintails at Loch Leven, and the Watchers Committee was subsequently formed to head this growing aspect of the work. Watchers schemes were not new; the first had been organised during the 1880s, but under the aegis of the RSPB, they became an organised and effective method of conservation. By 1914, the society was employing twenty-two watchers.

The idea of nature reserves, although catered for in the Charter, was still somewhat ahead of its time in the early 1900s, and the RSPB did not acquire land for some twenty-five years. But the intensity of work in the Society's formative years was tremendous. Remembering the almost wholly voluntary nature of the work, and the continuing full-blooded fight against the plumage trade, we see the RSPB nevertheless broadening its concerns to tackle the gamekeepers' pole-trap, the threat of oil pollution and the large numbers of deaths of migrating birds at lighthouses. At the head of this indomitable band was Winifred, Duchess of Portland, who was President of the Society from its inception in 1889 until her death in 1954. The RSPB, baby of a few nineteenth-century women, conceived out of emotion, was maturing into a pillar of strength in the twentieth-century conservation scene and was doing much to promote 'that now considerable body of opinion, the bird watchers' (Darling 1971).

On 12 January 1895, the National Trust (NT) for England and Wales was registered. It too became a colossus in the conservation world, concerned not only with the preservation of historic buildings but also with the conservation of nature and landscape. So much is it a part of everyday jargon that many people wrongly believe it to be a government department. 'The National Trust for Places of Historic Interest or Natural Beauty' had been nurtured for ten years in the minds of its makers. Canon Hardwick Rawnsley was a seasoned countryside campaigner. He had founded the Lake District Defence Society which took up its cause in 1876 when the Manchester Corporation dammed and raised the level of Thirlmere to allow piping of water to the city. In 1885 his efforts to retain public access in the Lake District brought him into contact with the other trust founders, Octavia Hill and Sir Robert Hunter. Both founder members of The Commons Preservation Society, they were deeply involved in the struggle for public access but

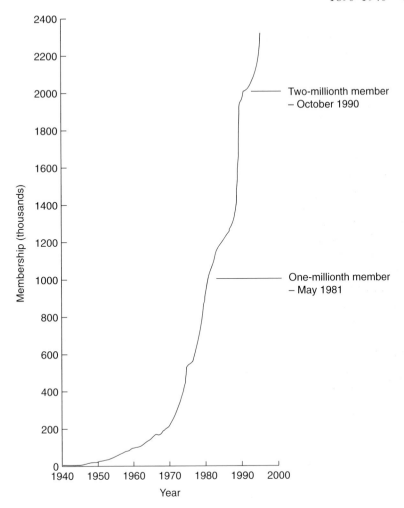

Figure 3.1 National Trust membership

sorely frustrated by that body's inability to buy land. The trust's aims were:

> to promote the permanent preservation, for the benefit of the Nation, of lands and tenements (including buildings) of beauty or historic interest; and, as regards land, to preserve (so far as practicable) their natural aspect, features, and animal and plant life.
>
> (National Trust Act 1907)

The trust did not receive a royal charter. Instead, the National Trust Act of 1907 became the foundation stone upon which its strength was built.

Plate 3.1   As the nineteenth century drew to a close, concern for the Lake District manifested itself in the formation of the National Trust, now the largest conservation organisation in Britain: Blea Tarn in the Lake District National Park. Photograph by Mike Williams. Courtesy of Countryside Commission

Hunter introduced the Bill, which passed through Parliament unopposed. Its backbone was a clause which allows the trust to declare its property inalienable. It cannot be sold, mortgaged or given away, although inalienable land may be leased with the permission of the Charity Commission. Nor can it easily be subject to compulsory purchase; by virtue of a 1946 Act, passed when compulsory purchase first became widespread, the trust has right of appeal to both Houses of Parliament against any efforts to purchase inalienable land. No other private landowner in the country has this right, and it has been an important point in encouraging owners both of land and of houses to place them in the hands of the trust. The earliest acquisition of land came in 1895 with a little less than 2 ha of cliff at

Dinas Oleu near Barmouth and, by 1910, thirteen sites of natural history importance were on the books, most notably Wicken Fen in Cambridgeshire. But the site selection was haphazard, always secondary to the acquisition of buildings, and opportunities were wasted through lack of capital and forward planning.

It was because of the continuing haphazard and cumbersome manner in which conservation was being practised that the last of the 'big three' influential societies was formed. The Society for the Promotion of Nature Reserves (SPNR), officially inaugurated on 16 May 1912, was the creation of Nathaniel Charles Rothschild. It was formed to fulfil a need and, over the years, it has altered its direction and its name on several occasions as new needs and new niches have been identified. Today, it is the Royal Society for Nature Conservation (RSNC) – the national association of Wildlife Trusts.

The need in 1912 was to place conservation on a more organised footing by preparing a shopping list of potential nature reserves for the benefit of organisations and individuals with the requisite funds. The need for reserves was still not accepted, and so meaningful surveys were non-existent. Even by the 1930s, when:

> issues such as national parks and rights of access to open country were built up ... as popular causes, the case for nature reserves received little publicity and generated no widespread interest. Instead, it remained an esoteric matter, viewed even by naturalists as a costly and impractical expedient only to be contemplated as a last resort when a unique spot was threatened by an improving farmer or speculative builder, and certainly no substitute for protective wildlife legislation.
>
> (Lowe and Goyder 1983)

It was not thought to be feasible to preserve wildlife on islands in the wider countryside, and the actions of esoteric individuals through the years had met, if not with open amusement, then with incomprehension. Imagine the disbelief of the local people, not to mention his own gamekeeper, when Charles Waterton, early in the nineteenth century, spent £9,000 on building a 5 km wall 2.5 m high around his Walton Hall estate in Wakefield to exclude sportsmen and their dogs. He declared the estate a nature reserve, erected nestboxes and forbade his gamekeeper to shoot owls and other predators.

There were a handful of reserves by the start of the present century. In 1888, for instance, the Breydon Society in East Anglia had bought Breydon Water and declared it a bird reserve. But such initiatives were rare and isolated. Rothschild was a quarter-century ahead of his time in seeing the need for an integrated string of reserves across the country. It has been said that, through his foresight, he 'invented' conservation as we know it. Both the name and the aims of the SPNR outlined the new direction:

- to collect and collate information as to areas of land in the United Kingdom which retain primitive conditions and contain rare and local species liable to extinction owing to building, drainage, disafforestation, or in consequence of the cupidity of collectors;
- to prepare schemes, showing which areas should be secured as nature reserves;
- to obtain such areas and, if thought desirable, to hand them over to the National Trust for Places of Historic Interest or Natural Beauty under such conditions as may be necessary;
- to preserve for posterity as a national possession some part at least of our native land, its faunal, floral and geographical features;
- to encourage the love of Nature, and to educate public opinion to a better knowledge of Nature Study.

The last was a universal aim of the movement at that time; the remainder formed a new spur from the mainstream of education and protective legislation. There was no intention that SPNR should become a landowner; it would promote reserves by attracting funds from various sources and placing the sites in suitable hands. Overtures were made to the wealthy, and Rothschild himself, often anonymously, put up funds for the purchase of sites. Blakeney Point, Ray Island, Wicken Fen and Woodwalton Fen were all acquired through his generosity.

Surveys were taken in hand and, by June 1914, ninety-eight sites had been covered. Increased pressure on land for farming at the start of the war intensified the urgency. In 1915, a provisional list of potential reserves was presented to the Board of Agriculture. It named 284 sites, covering all parts of Britain and Ireland. They were graded into three categories, their special interests were noted and details of ownership kept available. Any rare species were listed in separate sealed envelopes. In the following year, with a few additions, the list was confirmed. It is an accolade to Rothschild and his workers that the majority of the sites mentioned are still seen as prime places for wildlife today.

Unfortunately, the presentation of the list had no effect upon the government. It was buried under the weight of wartime matters and, by 1918, Rothschild was terminally ill. He died in 1923 and 'conservation in the modern sense of the word virtually ceased to exist in the United Kingdom' (Rothschild 1979). With the master gone, SPNR foundered. It remained in oblivion until the world caught up with Rothschild's ideas in the 1940s.

The First World War was followed, not by optimism for the future, but by discontent, apathy and inflation. The young conservation movement suffered, with much early momentum lost and new initiatives stifled by disinterest. It became almost impossible to awaken new appreciation for the natural world. Small wonder then that the movement remained too weak to enter the arena of land ownership. Funds were utterly lacking and public appeals could be confidently expected to fail. Rothschild had bought Woodwalton Fen to give to the National Trust but that body declined the liability of high management costs. With no takers, the owner had to hand it to his own SPNR, together with an endowment towards costs; yet this proved insufficient and the society was forced to sell Ray Island, another of Rothschild's gifts, in order to finance Woodwalton. Thus it became a reluctant landowner. It succeeded in shedding parts of Wicken Fen into the lap of the National Trust, which also agreed to take on Sharpham Moor in Somerset, on the condition that it would not have to finance its maintenance. It was inevitable that the few reserves already bought would be spoiled by lack of management.

LAWS AND LISTS

While the 'education' and 'reserves' arms of the movement continued to attract little real support, the 'protection law' lobby maintained its high success rates. The year that saw the start of the war and put an end to all hopes of a chain of early reserves nevertheless saw the first legal protection given to a British mammal (apart from game animals). Prior to 1914, the clubbing to death of seals, which we now view with disgust when practised by other nations, was a common pastime in Britain. In spite of this, the 1914 Grey Seals (Protection) Act, unlike the bird protection Acts, was based not upon emotion, but on a genuine conservationist concern that seals were facing extinction. It was known that they only produce one pup a year and the total population of grey seals was believed at this time to be no more than 500. This is now thought to have been a gross underestimate. Nevertheless, fishing interests were strong, and the Act, which introduced a close season between October and mid-December, was to run for a trial period of just five years. In fact it ran until it was replaced by a 1932 Act of the same name which prolonged the close season from 1 September to 31 December and extended measures to Northern Ireland. It had been found, in spite of the Act's glaring weaknesses, that the population had increased ninefold between 1914 and 1928 and yet fish stocks had apparently been unaffected. Grey seals have continued to thrive under the Act (p. 145) and Britain now boasts 60 per cent of the world population. Enormous 'rookeries' exist in the Scottish Islands, and new ones are being formed on other parts of our coast. In contrast, the population of the so-called common seal is about a quarter of the grey.

There was also concern for the badger. Poisoning was made illegal in England and Wales in 1911 and in Scotland a year later. Numbers then began a gradual climb but this did not last beyond 1945 (p. 147).

It was the birds that continued to gain most from the law, due to the unceasing efforts of the RSPB. Six years after taking up the cause against the pole-trap, it succeeded in having it outlawed by the 1904 Wild Birds Protection Act. A 1908 Act illegalised the teagle, a particularly nasty snare of baited hooks joined by strings which was put out to attract birds during hard weather. Bird lime and certain types of decoy were ousted under a 1925 Act.

But the RSPB's *cause celebre* remained the battle against the plumage trade until it was effectively ended with the 1921 Importation of Plumage (Prohibition) Act. The intervening years had not been easy. Opposition from vested interests had stalled an attempt in 1908 to pass a similar measure, until the war snuffed the life from it once and for all. Wartime trade restrictions, however, all but finished the plumage trade and took the pressure off the birds and the conservationists for a time. Yet the Board of Trade then chose 1917 in which to prohibit the importation of some feathers. The 1921 Act was the final extension that banned imports of all bird plumage except that of the ostrich and the eider duck. The millinery trade and its customers put up stinging opposition. Several attempts failed before the Act was passed and it is significant that the 'wearing of the ostrich' revived in the years that followed.

This and the 1925 Wild Birds Protection Act were the only two to reach the statute books during that decade. Four other bills failed, largely because the conservationists were trying to revamp the whole structure of bird protection law. Instead of the existing 'White List' arrangement, they sought a system based on a 'Black List'. The numerous Wild Bird Protection Acts by now in force were extremely complicated, confusing and, in effect, unworkable. Local orders had brought about a situation whereby the level of bird protection varied across county boundaries. If a bird was not specifically named in the White List of protected species, it was unprotected. Cumbersome enough at the best of times, this system broke down under the array of local and dialect names. (Jackson (1968) lists 4,840 names given to a total of just 329 birds.) The RSPB wanted a Black List arrangement, whereby birds which could be taken were specified and anything not mentioned was automatically protected. They argued the case for two Black Lists: one for birds to be taken only outside the breeding season and a second for birds which could be taken at any time of the year. It took the society ten years to get a government enquiry into its proposals set up in 1913. In spite of the war, the report was published in August 1919. Although it viewed the RSPB proposals favourably, it felt that the system was too complicated to invoke, again because of local and seasonal variations, and it advocated improvements in the existing legislation. On its recommendation, the Wild Birds Advisory Committees for England and Wales and for Scotland were set up in 1921.

But the RSPB still believed in the Black List system. Consequently it found itself opposing the four bird protection bills put forward by the Wild Birds Committees between 1923 and 1927. Thus it was that all four bills failed, mainly because of the conservationists.

The situation was soon recognised as being counter-productive. The Black List arrangement was too avant-garde, and the conservationists knew it. They could not agree even amongst themselves as to which species to show on which list. When the British Trust for Ornithology (BTO) was set up in 1933 as a research body (p. 54), this was one of the tasks it set out to tackle. Meanwhile, the RSPB accepted the need to seek improvements within the constraints of the existing regime. There was certainly much room for improvement. The trade in cagebirds had become so heavy that all possible steps to reduce it deserved careful consideration. Cagebird shows were popular and converted many to the pastime, thus providing an endless market for the suppliers. The official guide to the annual Crystal Palace show, billed as 'an innocent recreation', suggested that the way to ensure that your skylark would sing on the day was to keep it in darkness for four days beforehand and to feed it on whisky-impregnated worms. Alternatively, you could put out its eyes.

The 1933 Protection of Birds Act made it illegal to take, sell, offer or possess any one of sixty-six British species, while later steps added other birds to the list. This Act was the last bird protection measure for over twenty years, apart from certain steps taken to safeguard sporting interests (p. 51).

In Northern Ireland, there was a Wild Birds Protection Act of 1931, thanks largely to the recently formed Ulster Society for the Protection of Birds. Under this Act, a Wild Birds Advisory Committee for the province was organised. A further Act of 1950 embodied many of the committee's recommendations and, for a time, put Ulster ahead of mainland Britain in the legal protection of birds. Perhaps because there was no legacy of older law, a Black List system was introduced from the start. All birds, their nests and eggs were fully protected all year round unless they fell within one of three categories, which basically catered for the needs of the farmer and the sportsman. Various tools of the trade – the teagle, pole-trap, bird-lime and some decoys – were outlawed and the Minister of Home Affairs was empowered to declare nature reserves.

As well as leading the fight for improved legislation, the conservationists were involved in other efforts to protect our birds. The RSPB's Watchers' scheme continued, but it proved haphazard and expensive. Few prosecutions were taken, and its main effect had been to raise public awareness of the fact that some birds were protected by law. Those that knew it best – the egg collectors – nevertheless flourished under the auspices of the British Oological Association. So in 1936, the Association of Bird Watchers and Wardens (ABWW) was set up to guard the nesting sites of rare birds. Over sixty watchers performed duty in the first year. Within four years, the two associations, pressed by a mutual respect for the damage the other could do, were forging links in an effort to come to some agreement. Although the details were worked out, the war put paid to any hopes of a lasting settlement. On other fronts, the ABWW continued to do good work. Gamekeepers were paid for the successful breeding of a bird of prey on their estate. This was financed by members adopting a nesting site; 'adoption' is not such a new idea in conservation circles! In 1947, the association merged with the RSPB, and the wardening of nest sites and the payment of rewards to landowners remains current RSPB policy.

While the tangled web of bird laws continued to confuse, British flora remained unprotected. The early efforts at local by-law protection were sparse, piecemeal and thoroughly ineffective. Anybody anywhere could collect any plant in any quantity they wished. It was not until the 1920s that the conservationists took up the cause. The Botanical Exchange Club (p. 37) was revitalised and joined its voice to others calling for sensible measures. The SPNR, for all its languishing, ran a poster campaign exhorting people to pick sparingly, and Sheail (1976) tells of one landowner replacing its poster with his own notice forbidding all picking. The first conservationist appeals were broadcast on radio, particularly at bank holiday time. In the face of growing pressure, the County Councils' Association drew up a by-law in 1928 which banned the uprooting of ferns and wild flowers growing in any public place. Although under no compulsion to do so, twenty-five counties took the by-law on board. But if this measure had any effect at all upon the collectors and pickers, it was because they desisted voluntarily. Only covering public land as by-laws can, how was anyone to prove that the bunch of flowers in the collector's hands did not come from private land? Or from public land in a county where the by-law was not in force? And if the flowers were considered to be a weed, would it be wise to prosecute? The blanket protection of all species – the direction in which the bird protectionists were pushing – proved to be unworkable for plants. Schedules of specifically protected plants may have made the enforcement more practical, but these were not produced. Lists of particular rarities were published for certain counties by the Wild Plant Conservation Board but they only served to alert the collector. A new by-law of 1934 failed to improve the situation. In 1975, Perring could find no record of a successful prosecution under these by-laws. Britain's flora was to wait another forty years for any real legal protection.

SCIENCE AND SOCIETIES

But our flora was not without its champions. As the century moved on, a more scientific approach was employed in the study of natural history. Survey and research were deemed necessary and desirable. The number and range of societies inevitably grew in response to this widening need.

The Committee for the Study of British Vegetation, for instance, was formed in 1904, largely through the support of A.G. Tansley. Its brief was to co-ordinate the mapping of our native flora, which had begun in Scotland as early as the 1890s; to build up a photographic gallery of plants; and to make in-depth studies of certain areas. In April 1913, the committee expanded to become the British Ecological Society (BES) and it continues to flourish as such today. Its objects are 'to advance the education of the public and to advance and support research in the subject of ecology as a branch of natural science, and to disseminate the results of such research'. (For a full history, see Sheail 1987.)

Other long-running societies were formed in the early years of the century. E.K. Robinson founded the British Empire Naturalists' Association and its magazine, *Countryside*, in 1905. Both are still going strong, although the 'Empire' has been dropped from the name (BNA).

More conservation-minded from the start was the precursor of today's Fauna and Flora Preservation Society (FFPS). The 'Society for the Preservation of the Wild Fauna of the Empire' was the first in the field of overseas conservation. Launched by Edward North Buxton on 11 December 1903, many of its early supporters were both influential people and notable naturalists, and its membership reached 100 in the first year. Most were sportsmen, concerned for the future of their pastime – another sign of their vested interest in conservation. The FFPS works for the conservation of habitats and species, for protective legislation and for education, and it is involved in the captive breeding of certain rare species for release into the wild. For its first half-century, the interests of its founders ensured that most of its work went on overseas. Not until the 1950s do we see the FFPS moving more into the 'home market' as a result of growing concern for our domestic flora and fauna.

Fitter and Scott (1978) say of the original society: 'since the Empire at that time covered about a quarter of the surface of the globe, it was a fair start on internationalising the infant wildlife conservation movement'. This trend towards finding a wider base for conservation continued with the foundation in London, on 20 June 1922, of the International Committee (later Council) for Bird Preservation (ICBP). Dr T. Gilbert Pearson of the National Association of Audubon Societies in America was instrumental in getting it off the ground, and notable names on the British section included Lord Grey of Falloden, Frank Lemon, H.J. Massingham and Earl Buxton. The International Committee consisted of various national sections, each of these comprising up to eight representatives from the country's bird conservation bodies. A small staff was financed by these member bodies and, through bulletins and an international meeting every four years, ideas would be exchanged and the state of bird conservation in the world kept under constant review.

In its early days, the British section of ICBP was instrumental in drawing up the International Convention on Oil Pollution at Sea. Then, during the 1930s, came concern over the declining numbers of wildfowl in Europe. In 1936, a Wildfowl Inquiry Committee was set up and succeeded in bringing about The Wild Birds (Ducks and Geese) Protection Act 1939, which limited import of wildfowl during the close season. In 1947, the British section was closely involved in the formation of the International Wildfowl Research Institute (p. 89).

In 1988, with 300 member organisations in 100 countries, ICBP opened its doors to individual fee-paying members with the formation of its World Bird Club and Rare Bird Club. On 3 March 1993, it became BirdLife International, represented in Britain by the RSPB and in a total of 112 countries by twenty-one partners. Thus it retains its position as one of the most influential bodies in world-wide conservation.

The opposite trend in the 1920s was one of localisation which continues to prove, now as then, one of the greatest boons to the nature conservation movement in Britain. Local matters create interest, local problems cause concern, on a scale that far outstrips their national or international importance. Local feelings are generally protectionist, deep-rooted and very resolved. So it was a vital day in 1926 when S.H. Long and others founded the Norfolk Naturalists' Trust. Its aims were to protect suitable areas by establishing reserves; to prevent slaughter for fur, feather or skin; and to encourage the breeding of harmless, rare or beautiful birds (notice, no mention here of other types of wildlife!). Its immediate aim, and the catalyst which started the biggest movement within the world-wide conservation scene, was the need to preserve Cley marshes, 'the Mecca of ornithologists for the last 140 years' (Gooders 1967). Cley was bought within the first year. By 1941, the trust was managing fifteen reserves; by 1962, twenty-five.

The formation of the Norfolk Naturalists' Trust has been described as 'the most significant event between the wars' and 'a pioneer step' (Bonham-Carter 1971). Here was a trust as opposed to a society. A trust is a registered company and can own and lease land. A society, not being incorporated, is unable to do so unless it appoints trustees for the land. Nevertheless, British conservationists proved reluctant to follow the lead. Nature reserves were still not regarded highly enough to justify their expense. The RSPB acquired its first reserves in 1930 – Romney Marsh and Dengemarsh, and East Wood, Stalybridge – but the practicalities were different for those seeking to establish a local trust to purchase land. It was another twenty years before the example of Norfolk was followed by Yorkshire in 1946. Then came Lincolnshire (1948), Leicestershire, Cambridgeshire, and West Wales (1956) and West Midlands (1957). By 1958, the county trust movement was running 'like wildfire across England and Wales' (Fitter 1963) and SPNR had reviewed its role as a result (p. 88).

The Ulster Society for the Preservation of the Countryside (USPC) was established in 1937 in the face of fears over landscape despoliation, and subsequently moved into the field of conservation. Apart from the NT's Northern Ireland Committee, USPC is the oldest amenity society in the province.

There were no county trusts in Scotland but, late in the 1920s, the Association for the Preservation of Rural Scotland, itself only recently formed (p. 56), began to promote the idea of a National Trust. It came to fruition on 1 May 1931 with the official birth of 'The National Trust for Scotland for Places of Historic Interest or Natural Beauty' (NTS). Initially all efforts were applied to saving buildings but very soon the interest in amenity and national beauty began to bear fruit. The first countryside acquisition came in 1932 in the form of 600 ha of cliff and moorland on the island of Mull. In 1935, a place of great historical interest, natural beauty, amenity and conservation value was purchased – the magnificent Glencoe. The National Trust for Scotland had convincingly entered the field of countryside conservation.

It was not until 1935 that the trust gained official recognition in law. Having failed to win a royal charter, it succeeded in pushing through the National Trust for Scotland Order Confirmation Act. This outlined the aims of the trust, which included:

promoting the permanent preservation for the benefit of the nation of lands and buildings in Scotland of historic or national interest or natural beauty and ... as regards land for the preservation (so far as practicable) of their natural aspect and features and animal and plant life.

From an amenity point of view, the trust could:

maintain and manage or assist in the maintenance and management of lands as open spaces or as places of public resort [and] make all such provision as may be beneficial for the management of the property or desirable for the comfort or convenience of persons resorting to or using such property.

A further Act of 1938 granted inalienability to the trust's properties to bring it in line with its southern counterpart.

From a membership in 1932 of just sixty-four, NTS increased to 2,400 in 1950; 17,000 in 1960, 34,250 in 1970 and 130,000 in 1986. Its properties in 1986 totalled 39,000 ha. In spite of its modest growth and overriding concern with buildings, the trust became prominent in Scottish countryside matters and takes a leading role in nature conservation. 'I would say that the Scottish Trust now leads the world in the wholeness of its approach to environmental management' (Darling 1971).

OLD THREATS AND NEW

By the 1920s, the scientific aspects of conservation were becoming more evident. We cannot say that the majority of conservationists were yet acting in a scientific manner but, increasingly, the scientific content of their problems was prominent. As collectors and traders had been warned off, to some extent, by new laws and attitudes, more sinister threats were uncovered. Oil pollution was seen to be destroying seabirds in quite substantial numbers. The RSPB first passed comment in 1907 and, with the demise of the plumage trade in 1921, oil pollution became its primary concern. Within a year, the Oil in Navigable Waters Act 1922 prevented the discharge of oil within three miles (4.8 km) of the British coast. Moves towards an international agreement soon followed. In 1931, the society sank its teeth into an oil company which it prosecuted for discharging oil off the Pembrokeshire coast.

Another problem of scientific significance was the ogre of pesticides. This also had come to the notice of the naturalists in the first decade of the century with reports of garden birds killed by arsenic-based weed-killers and copper salts used to eliminate worms. Other vegetable-based substances such as pyrethrum and derris were available and safe, but being expensive to make, their use was never widespread.

The 1920s also saw an epidemic of Dutch elm disease, so called because the most serious studies into it were undertaken in Holland. Diseases of elm in Britain can be traced back to Neolithic times when, according to Rackham (1980), conditions were created that favoured such disease. There had been a bout in 1819–64, suitably bemoaned in printed records, but this had long been forgotten. When it came to notice again in 1927, it was hailed as a new threat to the British countryside and one upon which the wisdom of science must be brought to bear. In fact, all efforts to curtail the disease failed and, within ten years, 10–20 per cent of mature elms had been lost. But the recovery rate in Britain was found to be high so that, when the epidemic began to wane after 1936, complacency followed. No effective research was done into controlling elm disease and, when answers were needed again in the 1970s, the scientists could not provide them.

The scientific content of nature study and conservation nevertheless began to increase in view of this new type of concern. When the British Trust for Ornithology was formed in 1933, it was charged with the task of furthering our knowledge of birds. Early work concentrated on the drawing up of White and Black Lists (p. 49), on the effect of the little owl upon poultry and on that of the peregrine falcon upon homing pigeons. Such exercises required scientific survey and close observation. BTO continues its valuable work today, relying heavily on the records of amateur ornithologists all over the country. Many of its surveys are open-ended, such as the Common Bird Census which has provided valuable information on our much-loved garden species. The longest-running permanent survey is the Nest Record Scheme set up in 1939 as the Hatching and Fledging Inquiry, although great crested grebes have been surveyed from time to time since 1935 and grey herons since 1928, five years prior to the fledging of BTO itself. The best known enquiry is the Bird Ringing Scheme which has opened a window on the mystery of migration. Other pioneering initiatives have included the compilation of bird distribution atlases and an Ornithological Sites Register. Not surprisingly, BTO has been closely involved in the growth of bird obser-vatories in Britain. The first two were established in 1934 on Skokholm and the Isle of May.

Not all the efforts were for the birds. In 1926, RES founded its Conservation (Insect Protection) Committee and began efforts to re-establish the large copper butterfly at Woodwalton Fen (p. 106). Naturally, talk turned to other possible reintroductions and the topic became, as it remains, a highly controversial bone of contention amongst conservationists. Not only is it open to moral questioning but also it requires specialised scientific study (p. 267).

The oldest, if no longer the most serious threat to wildlife remained the still-popular blood sports. The practitioners were forming themselves into societies just as the conservationists were. The Gamekeepers' Association was set up in 1900; the Wildfowlers' Association of Great Britain and Ireland (WAGBI) in 1908. The two joined forces in 1975 and, six years later came to be known as the British Association for Shooting and Conservation (BASC). Fox-hunting and hare-coursing remained popular. Gamekeepers maintained their battle against the predators. Wild cats and pine martens reached their lowest ebb. Polecats were threatened; over 100 were being killed annually in central Wales alone. Early in the 1930s, the disease 'strongylosis' decimated partridge populations; in desperation, the gamekeepers stepped up their control of birds of prey. Adverse comment fermented into outright opposition and many conservationists became openly hostile towards the gamekeepers and sportsmen. They, in turn, reinforced their position with the formation of the British Field Sports Society (BFSS).

The field of battle was set and emotions were running high. Both sides realised that heavy losses were the most likely outcome from such entrenched positions. Perhaps science could be brought to bear, although upon which side it would fall was grossly unclear. One side was very aware that populations of prey and predator must remain in careful balance, but was unclear on how this should be achieved; the other knew that our wildlife had survived centuries of field sports. For the first camp, bodies such as the RSPB and the BTO began research into the prey/predator relationship. In the second camp:

> it was beginning to be realised that predators and prey had co-existed happily for thousands of years before man began protecting game. It was beginning to be realised that indiscriminate destruction of predators was a short-sighted policy, leading to disease and financial loss. More and more shooting-men began to accept the view that to maintain a healthy stock of game it was necessary also to maintain a stock of predators; that reduction rather than destruction was the correct policy. Some landowners began to afford protection to predators, at least to the less common, on their ground. For example, a number of owners and tenants of properties in Sutherland in 1939 agreed between themselves, without prompting from any society interested in preservation, not to trap or shoot the pine marten. No longer, over much of the

country, was the peregrine falcon shot at sight. The golden eagle and the wild cat began to increase in numbers and even to extend their range. There was a movement towards some sort of balance.

Had not the war intervened I verily believe that a workable balance between all the various interests might have been achieved within a few years to the great benefit of the countryside.

(Vesey-Fitzgerald 1969)

But the war that scuppered Europe also put paid to those early moves onto common ground. It left both camps floundering in the no man's land between their respective standpoints. Although treaties have been signed and many battles avoided, the uneasy peace is still too frequently marred by allegation and bad practice. Poisoning and, to a lesser extent, shooting and trapping are known still to be rife, particularly on the large estates. A total of 1,166 reports of illegally poisoned birds of prey during the 1980s must represent the merest tip of a very large iceberg. In 1994, the RSPB got to hear of 229 incidents of raptor persecution, 170 of them in Scotland.

The greatest threat of all to wildlife during the 1920s and 1930s came from the changes in land use that were happening at an unprecedented rate. It is a problem that brings us right into the realm of modern conservation. The increased use of land for farming during the war was mentioned as a factor in pushing through SPNR's survey of potential reserves. But the loudest noises were being made, not on behalf of our wildlife, but for our farmland; the greatest concern was over its loss to urban development. That loss was indeed staggering. The British population, which had been 70 per cent rural 100 years earlier, was now 80 per cent urban.

Between 1920 and 1940, no fewer than four million houses were built. The yearly net loss to concrete more than doubled in the late 1920s, and the annual loss of over 25,000 ha in each and every year of the 1930s has never been approached since. The word 'suburbia' has an ugly ring to it these days; in the 1930s, suburbia was the place to be. But the sprawl was seen as a terrible menace to the landscape and 'probably did more than anything else to trigger the alarm of conservationists about the countryside' (Adams 1986). At that time, ironically, the alarm was for the future of our farmland.

In 1926, the Council for the Preservation of Rural England (CPRE) was born to act as a watchdog over urban development and to seek the best use of land from an amenity point of view. There rapidly followed the Association for the Preservation of Rural Scotland (APRS) and the Council for the Preservation of Rural Wales (CPRW). The musty idea of 'Preservation' was replaced in their names by 'Protection' in 1970, but their aims remain the same. Like their urban equivalent, the Civic Trust – set up in 1957 when towns apparently became worthy of protection –

they act as umbrella societies, co-ordinating the activities of local amenity groups and encouraging their campaigns for sound environmental decisions in the planning world. They also promote legislation for the protection of the countryside and will advise landowners on the amenity aspects of their property. Twenty-two existing bodies joined CPRE at birth and helped it to grow into an influential and well-respected organisation. It was the first association to set up local branches with a view to local action. It runs the 'best-kept village' competition and its local view of events has helped it to many successes in local planning matters. There are fifty county branches of CPRE with a combined individual and group membership of 45,000. The CPRE and its sister organisations are registered charities, relying on donations and subscriptions to keep them independent of official bodies.

One such body with which CPRE came into early conflict had nothing whatsoever to do with urban development. The Forestry Commission (FC) was set up in 1919 by the Forestry Act after the war had highlighted short-comings in our timber reserves and production, as it had in food production. The UK percentage of forested land was the lowest in Europe, at 3 per cent or just 700,000 ha. The commission's brief was to improve the supply, production and reserves of timber both by its own operations and by promoting the idea of forestry to private interests. Above all, it was to be profitable, and it was this requirement that brought it into conflict with other countryside interests. Initially, the FC was to aim for new plantations totalling 710,000 ha in England, Wales and Scotland (it has no responsibilities in Northern Ireland). It set about the task with alacrity. In the first year, 19,000 ha were bought and 700 ha were planted. By its tenth birthday, there were 56,000 ha of new plantations and, by 1939, 172,000 ha. But such enthusiasm to become profitable, coupled with the inevitable teething troubles of a new concern, encouraged poor forestry techniques, the most notable to the eye and the landscape being symmetrical blanket plantations of conifers. It was a policy which rapidly brought the FC a bad name in amenity circles. That name has lived with it to this day in spite of policy changes and much good work towards the conservation and enhancement of our landscape.

Aware, even at this stage, of the need to improve its image, the FC convened a joint committee with CPRE in 1935 and began the long haul to become accepted and respected by amenity interests. In deference to the growing demand for public access to the countryside, it opened its first Forest Park in 1936, fifteen years before the first National Park. Taking in 14,000 ha in Argyllshire, it was soon equipped with visitors' hostels, camping sites, an information centre and forest trails, backed up by a guidebook. So successful was the idea, that Snowdonia (8,200 ha) and the Forest of Dean (9,200 ha) were designated by the end of the decade. Northern Ireland's Forestry Service followed the lead by establishing the Tollymore Forest Park in 1955.

After consultation and some compromise, the committee also came to an agreement to leave parts of the Lake District unforested. The clearing of this small hurdle was 'highly significant for it established liaison between the voluntary bodies and a statutory body on the future use and management of an extensive area of countryside' (Sheail 1976).

But let us have no illusions. None of these good works was performed on behalf of our wildlife; the pressure was coming from the 'access to the countryside' lobby. It may have been that wildlife would have benefited from even wider afforestation and the turning back of those disruptive feet. Stamp (1969) hazards the suggestion that:

> systematic forestry favours wild life conservation operations, so that while some groups may suffer temporary eclipse, others have a chance. A mature oakwood offers a single monotonous environment to both plant and animal life; it is relatively dull. The introduction of other forest trees, including a range of conifers, undoubtedly creates a range of environments and in consequence more varied wild life.

It is not a suggestion that I should like to have made. Oakwoods are monotonous neither to the sensitive human being nor to our wildlife, and they support a more varied flora and fauna than almost any other habitat. But in contrast to the bleak open tops of much of the Lake District, might not the varied habitat of a commission forest have proved more beneficial to wildlife? Nobody was arguing nature conservation.

NO ACCESS

Access to the countryside was the clarion call of the 1930s. To the trains and bicycles of the nineteenth century had now been added cheap motorcars, motorcycles and omnibuses. People could take to the countryside for day trips and holidays, to hike or to camp, to look and listen and take in the air. As the decade moved on and their demands became greater, the amenity lobby grew increasingly organised and vociferous. G.M. Trevelyan, in a lecture delivered in 1931, voiced the fears of many; that rapid deterioration of rural England had been rampant for over a century; that, for economic reasons, new beauty was not being developed, and that preservation was the only hope for the future. Not surprisingly, he was a leading light in the new amenity movement. He was President of the Youth Hostels Association for England and Wales (YHA), formed in 1930 'to help all ... to a greater knowledge, love and care of the countryside'. Similar associations were set up in Scotland and Northern Ireland in 1931, and the Ramblers' Association was formed in 1935 from the combined forces of the early access groups (p. 26).

It was in 1929 that CPRE, supported by its sister organisations, launched the long campaign for the establishment of National Parks. The Addison Committee was formed to consider the desirability and feasibility of such parks for preserving the natural landscape and wildlife and improving facilities for recreation. If we look at these two aspects, they quickly show themselves to be divergent. People, particularly when out for recreation, tend to disrupt the landscape and disturb the wildlife. The 'countrysiders' had failed to demonstrate just what it was they wanted, not for the first time and certainly not for the last. The 1931 Addison report did come out in favour of National Parks on both counts but, in practical terms, nothing immediate could be done.

Although Addison had considered the preservation of wildlife, the conservationists did not take up the banner of the National Parks. On the other hand, the amenity lobby was well aware of the potential advantage to their cause. They knew what they wanted and set out to force the politicians from their state of inertia. There were the organised 'trespass hikes' which made their way into the history books. Police and landowners found themselves skirmishing on the high moors with determined walkers. A mass trespass of 24 April 1932 on Kinder Scout in the Peak District ended with five men being imprisoned by Derby Assizes. Then there was a concerted offensive to have the Parks established during George V's Jubilee year of 1935. When this failed, CPRE and its counterparts formed a joint Standing Committee on National Parks. This attracted much support from the broad spectrum of outdoor interests and there were strident demands for an official National Parks Commission to consider the proposals. The Pennine Way Association was set up in 1938 to press for the opening of a long-distance footpath. The idea of coastal paths was also being floated.

By the end of the decade, the politicians felt that they had to be seen to be doing something. Thus it was that they passed the so-called Access To Mountains Act of 1939. Such a sham was it that many supporters of the original bill had to back down. Under the Act, landowners and local authorities would have had first to make application to the Ministry of Agriculture for their land to be 'specified'. The weight of work which this would have brought down upon the shoulders of the ministry was such that the system collapsed before the Act could be implemented, as planned, on the first day of 1940. The landowners breathed sighs of relief, and so did the hikers and ramblers. The Act would have made trespass a criminal offence, for the first time in British legal history, and the means of gaining legal access were to have been so convoluted as to be totally impractical. The hiker, in effect, would have been driven further from the moors or, as a trespasser, closer to the courts and a criminal record. Such was the depth of failure of the first Act to promote public access to the British countryside.

4 1940s: National parks and nature reserves

NATIONAL PARKS

It was another full decade before any steps promoting access to the countryside reached the statute books. The intervening war has often been blamed for the delay, but in reality the Act, when it came, was a direct result of the war. For during those dark years, a whole new concept washed over Britain: the concept of a brave new world, of a land fit for heroes. All that was best about Britain would be made available to the masses. The landscape would be theirs to enjoy, returned into the hands of the common man from those who had annexed it over the years. The leisure to enjoy it and the freedom of the open spaces would maintain the nation in physical and mental health, after the black horrors of war. The country-side took on a rosy hue, just as it had when industrialisation and when colonialism had exiled so many from it, but this time the current of opinion was stronger. The first concrete result of official resolve to thank the British people for victory in war was the National Land Fund, established in 1946. Under this scheme, surplus wartime supplies were sold off to pay the Inland Revenue for land or historic buildings offered to it in lieu of death duties. The property thus acquired by the Fund was then given to the NT for the benefit of the people. In 1953, the scheme was extended to allow for the acceptance of chattels. The NT was resolute in remaining independent from central government, and this scheme was preferred to any which involved the award of a regular grant. By 1977, it had accepted through the fund, fifty buildings and 44,000 ha of land, worth £6 million. But it still remained wary of relying on government and so the fund was reconstituted as the National Heritage Memorial Fund, with independent trustees and wider terms of reference.

Access to and care of land in the hands of the NT posed no problems. In the wider countryside, a system of planning controls was part of the new vision, to ensure that Britain's land would remain forever unmarred by inappropriate development. Longer holidays, higher pay and greater mobility were the rewards awaiting the people of this brave new world, to enable them to reap the very best of their newly won heritage. Hugh

Dalton, both Chancellor of the Exchequer and President of the Ramblers' Association, summed up the combined view:

> We must, in the lifetime of this Parliament, place on the statute book a great measure of liberation, freeing for the health and enjoyment of the people what so long has been monopolised for the few.

An internal memorandum of the Ministry of Health (quoted by Sheail 1976) demonstrates, even through the heavy language, the impressions within official circles of the new current of opinion:

> the course of events has led to the conclusion that the Government will be exposed to serious criticism and discredit if a purely negative reply continues to be given to the large body of opinion in favour of definite action for the preservation of the countryside. The National Parks appear to provide the best opportunity of making a gesture to indicate the reality of the Government's interest in the problem of preservation.

But for all the promising noises made by the politicians, the main drive came, as ever, from the voluntary bodies. It was they who identified the need and put forward the proposals to fulfil them. A YHA National Council motion of 1944 read:

> The Association is of the opinion that any satisfactory solution of the problem must be based on the provision of National Parks; the establishment of a Footpaths Commission to record Rights of Way and settle disputed cases; the creation of long distance paths ... and improved legislation providing access to uncultivated mountains and moorland.

By this time, the campaign of action had become a long and drawn-out affair, characterised by committees and their reports. The 1931 report of the Addison Committee saw National Parks as a defence in the countryside against haphazard development and an opportunity to widen public access to open spaces. 'Nature Sanctuaries' both within and without the parks were proposed for the protection of wildlife. But nature reserves were not part of the picture imagined by the access lobby, which was quite rightly sceptical of suggestions that public access should be mixed with nature preservation. They were quick to realise that nature reserves would not be open to free public access; far from it. So the proposals developed along two separate lines: recreational parks on the one hand and nature reserves on the other or, at least, wildlife havens within wider National Parks. For political and economic reasons, the recommendations in the Addison report were never instigated. But already a conflict of views was evident between those who wished to keep nature reserves totally separate from access areas and those who felt that the voluntary movement could only stand when united in its campaign.

When the Standing Committee on National Parks (p. 59) was set up by the voluntary bodies on 5 May 1936, it attempted to co-ordinate all interests – recreation, amenity, and nature conservation – in a concerted effort to force the hands of the politicians. One of its leading lights was John Dower. Through his White Paper of 1945 he, more than any other individual, influenced the popular vision of what National Parks should be. He firmly believed that all the outdoor interests should pull together if the campaign for the National Parks was to have any hope of success; that recreation and conservation must come together for the common good. It is a sentiment that should be remembered more often than it is fifty years on. Nobody, least of all John Dower, doubted that public access interfered with nature conservation, but in the face of other threats to the landscape, a common purpose had to be displayed by the expanding countryside movement. In his vision, he saw nature conservation as one of the recreational facilities to be offered in a National Park and he understood that public access to our wildlife and landscape was the surest encouragement to its conservation.

In July 1938, the Standing Committee published *The Case for National Parks in Great Britain*. It defined a National Park as 'an extensive district of beautiful wilder landscape, strictly preserved in its natural

---

John Dower's proposed twenty National Parks would have covered one-third of the UK:

England	Cheviots and Roman Wall
	Lake District
	Northern Pennines
	Mid Pennines
	South Pennines
	Cleveland Hills
	Norfolk Broads
	New Forest
	Dartmoor
	Exmoor
	Coast of South West Peninsula
Scotland	North West Highlands and Islands
	Central Highlands
	Argyll Highlands and Islands
	Southern Uplands
Wales	Snowdonia
	Mid-Wales
	Brecon Beacons and Black Mountains
	Forest of Dean and Wye Valley
	Pembrokeshire Coast

aspect and kept or made widely accessible for public enjoyment and open-air recreation, including particularly cross-country walking, while continued in its traditional farming use'. The report pointed out that Britain was lagging behind other countries in the launching of National Parks and that threats to potential park areas were increasing daily. It outlined a proposed policy on National Parks based upon a National Parks Commission for England and Wales, and another for Scotland. Their duties would be:

- the selection of National Park Areas;
- guidance and co-ordination of Local Authorities;
- co-ordination of government departments and other statutory bodies;
- allocation of government grants;
- assistance in the supervision and management of National Parks;
- development of policy and guidance of public opinion;
- a 'watching brief' over potential National Park areas.

John Dower, in a draft National Parks Bill in 1939, saw the proposed commissions mainly as planning authorities with a brief to appoint a committee to each park, to designate nature reserves within the parks, and to encourage public access and compensate landowners as necessary. It was widely recognised that the Town and Country Planning Act of 1932 was doing very little to preserve potential parks from damage since so many threats fell outside the scope of the Act.

In spite of official hints to the contrary, funds were not forthcoming. More weighty matters preyed upon political minds and nothing was done before war broke out.

With the war came the hopes of a brave new tomorrow. Future Britain would need to be a place worth fighting for. More control must be exercised over urban and industrial development; more thought given to the preservation of nature and the open spaces. At a time of full employment, industry lost some of its attraction and the widespread use of country areas as factory sites was called into question. The Uthwatt Committee of 1940 was set up to review planning matters and introduced interim controls in July 1941. The Scott Committee of October 1941 was charged with looking at land use in rural areas, considering not only economic and employment factors but also the welfare of rural communities and the retention of their amenities. With these two committees, we see the official view moving perceptibly towards countryside conservation.

The voluntary bodies were quick to notice the signs and stepped up the pressure with renewed vigour. Demands and proposals came from all quarters of the movement. But with them came the conflict between the two camps of recreation and conservation. It was imperative that efforts were made to reach compromise over proposals which could be presented as the common aims of a united movement. But it was far from certain that those efforts would succeed.

The Case for National Parks had already reviewed the problem. The most effective nature reserves would be those surrounded by another band of protected land; that is, reserves within a National Park. It was essential that such reserves should exist but also that general measures of wildlife protection should be practised throughout all the parks.

> Broadly speaking, however, Nature Reserves are not in themselves National Parks, and many of those most urgently required for the preservation of rare and localised species are not in areas suitable for National Park use. The establishment and control of Nature Reserves requires separate and distinct measures, which the National Park system should do all it can to assist, but for which it cannot be solely responsible.

The idea of nature reserves had come of age. The fear that they would become honeypots for collectors was eclipsed by the reality of their ruination by increased wartime food production and urban and industrial development. The doubts as to their usefulness as isolated plots fell away before the realisation of their value as undisturbed havens. The fact that the majority of people still needed to be taught how to behave correctly towards their natural heritage was seen, not as a reason to delay setting up reserves, but as a prime objective for doing so.

On 5 June 1941, the RSPB and the SPNR convened a 'conference on nature preservation in post-war reconstruction'. Over thirty voluntary bodies and a number of local authorities were represented. The conference set up the Nature Reserves Investigation Committee (NRIC). In March 1943, NRIC published a memorandum, *Nature Conservation in Great Britain*, which listed sixty-one reserves already in existence. Shortly afterwards, it began a survey of all potential nature reserves within the likely National Park areas. Amateur naturalists were encouraged to submit local sites for consideration but the main input came from BES. In listing sixty-seven separate habitat types, it put forward the idea of forty-nine 'national habitat reserves' to include at least one example of each type, together with thirty-three 'scheduled areas' where development ought to be restricted on behalf of wildlife. The eventual NRIC proposals, published in December 1945, listed forty-seven potential National Nature Reserves and twenty-five 'conservation areas'. In total, these would have accounted for an area somewhat less than 3 per cent of England and Wales.

The APRS had been involved in the Standing Committee from the start. But north of the border, the terrain and the relative lack of pressure upon it, together with the different concerns of the rural communities, combined to make it a very different ball game. Any policy on National Parks would have to be geared specifically to the Scottish situation. So, in 1942, the APRS formed the Scottish Council for National Parks. The council did what it could to keep the case for National Parks to the fore, but the need was not pressing and subsequent development merely reflected the efforts being made further south. Likewise, very little was done to identify potential nature reserves. The Zoological Society of Edinburgh, founded in 1909 to 'study and preserve wild animals', had spent a quarter of a century setting up the National Zoo! In 1938, it organised a Scottish Nature Reserves Committee in conjunction with the APRS, the National Trust for Scotland and the Forestry Commission. But this remained totally ineffective, in spite of prompting from the Council for National Parks, so that when the Ritchie Committee (p. 75) came to tackle the job eight years later, it had to start from scratch.

Given the greater pressures in England and Wales, things were moving along in a much more definite manner. The Field Studies Council (FSC) was set up in 1943 as the Council for the Promotion of Field Studies, in the wake of growing interest in nature reserves for their educational potential. The idea fitted in very well with the vision of a brave new tomorrow and, although the original aim was to encourage the amateur naturalist, it coincided with a change in outlook and syllabus in the schools and colleges which subsequently provided most of the FSC's clientele. Flatford Mill, their first field centre, opened in 1946 and was followed in 1947 by Dale Fort, Juniper Hill and Malham Tarn.

The educational value of reserves was only one aspect of their perceived potential. They could also be used for scientific research, for economic research, as a recreational amenity and, of course, as havens for wildlife. All were seen as valid aspirations, even though they would rarely make comfortable bedfellows.

> We must have several attitudes in mind, and we must be careful that our special fancies for any one of these attitudes does not lead us to neglect the weight of the others. Our attitudes must be biological, scientific, economic, and aesthetic.
>
> (Fisher 1940)

Meanwhile, both the Uthwatt and the Scott Committees had come out in favour of radical planning controls. The Scott report of August 1942 called for a balance between urban/industrial use and agricultural use and referred to the establishment of a National Parks Authority as 'long overdue'. Specifically, the report revolved around five themes:

- the setting up of the National Parks;
- the consideration of the coastline of England and Wales as a single entity within any National Parks programme;
- the establishment of nature reserves as distinct from National Parks;
- the establishment of a 'footpaths commission' to survey and record public footpaths, to enquire into the legal situation and settle disputes, and to press for new paths such as the Pennine Way;
- the registration of common land, together with a review of the commoners' rights and a study of future maintenance and protection of public access.

Following the publication of the Scott report, the government appointed John Dower to review the practicalities of these proposals as they related to public recreation, nature conservation and the preservation of buildings. His report of May 1945 has been described as 'a masterly document – imaginatively conceived and expertly presented . . . now acknowledged as the basic blueprint of recreational planning in the countryside' (Bonham-Carter 1971). Its vision of a National Park was broadly that of the earlier report:

> an extensive area of beautiful and relatively wild country in which, for the nation's benefit and by appropriate national decision and action:
>
> – the characteristic landscape beauty is strictly preserved
> – access and facilities for public open-air enjoyment are amply provided
> – wildlife and buildings and places of architectural and historic interest are suitably protected, while
> – established farming use is effectively maintained.

The report listed twenty-two potential parks covering over 2 million ha. With foresight, Dower warned that the first ten should be set up by 1950, lest tourist pressure become too great upon those first notified. Given the

Plate 4.1 Dower's National Park vision was of a working landscape in which the best aspects of scenery, wildlife and historic interest were preserved for public enjoyment: inland from County Gate, Exmoor National Park. Courtesy of Countryside Commission

pressure on our parks today, would Dower now have seen the time as right to designate the remaining areas on his list?

The government of the day was not blessed with Dower's sense of urgency nor, it seems, with his foresight. In the face of opposition from vested interests to the likely power of a National Parks Commission, not to mention the expense of setting it up, the Ministry of Town and Country Planning saw fit to convene yet another committee. A vital opportunity had been lost. This committee, set up in July 1945, was chaired by Sir Arthur Hobhouse. Its brief, rather lamely, was to review the case for and against National Parks and access to the countryside and to consider the legal machinery needed to instigate them. Hobhouse, himself inevitably embroiled in the mesh of the committee mentality, organised two sub-committees. In doing so, he emphasised the existence of two camps. The Wild Life Conservation Special Committee, chaired by J.S. Huxley, was left to consider nature reserves (p. 70). Hobhouse himself oversaw the Footpaths and Access sub-committee along with the main National Parks board. The latter reported in July 1947 with the 'footpaths' report following two months later.

It is clear from his findings that Hobhouse draw heavily upon Dower's ideas and that he accepted the need for a National Parks Commission to oversea their running. But he did not share Dower's incentive to get them established quickly, nor did he believe in giving them a high level of independence. He recommended that the commission should be made responsible to the Ministry of Town and Country Planning. He advocated twelve National Parks, covering somewhat less than 1.5 million ha; they included the ten we have today, together with the Norfolk Broads and the South Downs. He also recommended fifty-two conservation areas including the Brecklands of East Anglia, the Cotswolds, the Malverns, and fourteen separate stretches of coastline.

Had Dower's recommendations of 1945 been acted upon when they were published, with the sense of urgency felt by their author, things might have been very different. As it was, the Hobhouse findings were published after the passing, at a substantial cost of £300 million, of the 1947 Town and Country Planning Act. The Act had given county councils wide powers to control development and they were not about to relinquish any measure of these powers to a National Parks Commission. The planning mechanism was new and novel, untried and therefore unfaulted. No doubts could be entertained about its expected efficiency; it was the unquestioned harbinger of sound, sensible development and the saviour of our natural landscape. The Act included many measures to help protect the countryside, such as the much heralded and sacrosanct Green Belts. Any differences between developer and amenity interests could always be ironed out at public inquiry if the planning authority felt it necessary. The National Parks Commission then would need to hold no planning powers but to act purely as an advisory body to county councils on planning matters.

While there is no doubt that high hopes for the new planning machinery were genuinely held, the voluntary bodies were quick to voice their concern over its early shortcomings, as they have been forced to do ever since. The YHA set up Regional Countryside Committees in December 1947 to watch over proposed changes in land use, to protect existing footpaths and to encourage new legislation on public access to the countryside. The Committees were careful, too, to encourage the good behaviour of YHA members when out in the country. The Ramblers' Association was busy forcing improvements in the access provisions of the National Parks and Access to the Countryside Bill. It succeeded in putting greater responsibility upon county councils to survey and publish 'access land' and to report on their intentions to open up further areas.

The voluntary bodies, co-ordinated by CPRE were also deeply committed to recovering land from the clutches of the armed forces now that hostilities had ended. In 1946, the services not only controlled over 400,000 ha but were calling for 1.2 million more, and they had the right to take it without public inquiry. Public rights of way, withdrawn during the war, remained closed. A report in the *Birmingham Mail* of November 1947, quoted by Coburn (1950), refers to the coastline:

> Erosion continues, accommodation is inadequate, shacks and shanties deface many of the finest bays and headlands, military barriers and barbed wire still litter the shores and dunes, the coastal path exists only in bits and pieces, and many of the finest stretches are earmarked for exclusive and permanent use by the Services.

The same situation of occupied territory persisted in many mountain and moorland areas. Public opinion began to run high now that the promise of a new start had to be fulfilled. The government opened an enquiry on seven test regions and one or two small areas were returned to public use. But the forces had dug well into their trenches. A joint YHA/Ramblers' Association petition of one hundred thousand names, presented to Parliament in December 1947, was parried in the same month with a White Paper on Service Department needs. Progress in getting the land returned to the people promised to be very slow. By and large, however, our wildlife was better served by the generally passive regime of the Services than it would have been under the attack of public access.

In Scotland, the National Parks movement was taking a different tack. Pressures in the countryside were not nearly as great, either on the ground or from the amenity bodies. The committee working under Sir Douglas Ramsay, on behalf of the Scottish Council for National Parks, reported in 1945 and 1947. *National Parks and The Conservation of Nature in Scotland* suggested five 'priority' and three 'reserve' National Parks covering less than 675,000 ha. With a land area exceeding half that of England and Wales, Scotland stood to gain far less in the way of National Parks.

Ramsay's recommended National Parks for Scotland were:

Priority	Loch Torridon, Little Loch Broom, Loch Maree Glen Affric, Glen Cannich, Strathfarrar The Cairngorms Ben Nevis, Glencoe, Black Mount Loch Lomond and the Trossachs
Reserve	Moidart, Morar and Knoydart Ben Lawers, Schiehallion, Glen Lyon St. Mary's Loch

Three nature reserves were also recommended:
Black Wood of Rannoch
Moriston
The lower reaches of the Garry.

Ramsay, like Hobhouse, saw a ten year time span as not unreasonable for the establishment of the first National Parks, in this case the five 'priority' regions. As it happened, they never were established. Initially, hurdles were met because Ramsay had advocated State purchase of the parks. Landowning interests erected barriers which proved insurmountable. But the main reason for the failure of the National Parks dream in Scotland was its lack of urgency, its lack of relevance. At the time, there was no shortage of open land available to resident and visitor alike. To 'close off' vast areas purely for recreation would be to threaten the viability, the very existence of the local communities. Not that Ramsay had recommended that course of action. He envisaged a National Parks Commission in Scotland providing grants to encourage tourism, agriculture and light industry to the parks. But this did not accord with the general ideal:

> Within internationally accepted national park standards, conservation must always take precedence over recreation and other land uses. This was seen as lacking the flexibility needed to develop land management policies which seek to conserve traditional qualities and national and regional characteristics.

> (CCS 1986)

The result was, quite simply, no National Parks for Scotland. Ramsay's five 'priority' parks became 'National Park Direction Areas' for which all planning applications had to be referred to the Scottish Office. And that was as far as it went. The greatest loss, perhaps, is that, without a National Parks Commission, there was no central body to oversee countryside matters in Scotland; no agency to lobby government or to co-ordinate public opposition to damaging development. That ball remained in the court of the voluntary movement until the setting up of the Countryside Commission for Scotland (CCS) in 1968.

NATURE RESERVES

Running parallel to the National Parks movement was the campaign for a series of National Nature Reserves (NNRs). In spite of Dower's efforts to unite the two, they developed separately. Dower and a handful of others feared that two sets of machinery would prove costly and cumbersome and less likely to succeed. He wanted to see National Parks Authorities running reserves within their regions and local authorities, or even voluntary bodies, administering those without.

But the naturalists and conservationists could not come together with the amenity lobby. The nature reserves they sought would not prosper in conjunction with public access. Furthermore, most would fall outside the boundaries of the National Parks. They wished to see some uniformity of administration by means of a central body, not the vagaries of disjointed local stewardship. There were growing demands for an official ecological research body to study wildlife and its management and to maintain nature reserves. For, while many were exhorting the praises of reserves, nobody had any practical experience of the intricate art of managing them.

It was no surprise then that Hobhouse appointed his sub-committee under J.S. Huxley to consider the question of conservation in isolation from public amenity; nor that this committee consolidated that isolation once and for all. In fact it was a novel and baffling experience to many of its members to have to mix even with other naturalists.

> It was only now that naturalists from different specialisms found themselves compelled together to examine specific areas in the round and to try to agree what to do about them. The discipline was salutary, and the influence of such a profound intelligence and far-reaching sympathy as A.G. Tansley's brought home the parochialism and narrowness of most individual and even corporate concern with nature.
>
> (Nicholson 1970)

It was the report of Huxley's Wild Life Conservation Special Committee that rationalised the arguments and outlined the foundations on which present British nature conservation policy is based. *Conservation of Nature in England and Wales*, published in July 1947, was a vital and stimulating piece of work which considered both the means of selecting nature reserves and their subsequent administration.

> Its treatment of the reasons for conservation, and its analysis of ways and means, are probably still the most lucid and straightforward

statement of what conservation is about even though (or perhaps because) it was not in civil service prose.

(Adams 1986)

Its vision of NNRs is seen by Mabey (1980) as 'the single most important influence on the development of nature conservation in Britain'.

It is impressive that the original assessment of the great majority of sites proposed as NNRs remains unchanged in the present review. It is, indeed, remarkable how consistently unanimous a view is to be found amongst experienced field ecologists and naturalists about the outstanding merits of many key sites.

(NCC 1977)

Huxley (1947), balancing the divergent needs of a reserve as wildlife haven, laboratory and classroom, recreational amenity and economic asset, concluded that selection should be based on the scientific value of a site in relation to our overall resources of wildlife and habitat. The series of reserves should:

preserve and maintain ... places which can be regarded as reservoirs for the main types of community and kinds of plants and animals represented in this country, both common and rare, typical and unusual, as well as places which contain physical features of special or outstanding interest .... Considered as a single system, the reserves should comprise as large a sample as possible of all the many different groups of living organisms, indigenous or established in this country as part of its natural flora and fauna.

The administration of the reserves should be invested in one body, totally separate from any National Parks Commission and including scientists and researchers. Its duties would be to:

- select, acquire and manage reserves;
- undertake survey and research work;
- be a central information bureau and advisory service;
- encourage educational work;
- initiate new legislation and ensure enforcement of existing conservation laws.

Specifically, Huxley was almost exactly in line with the earlier findings of NRIC in recommending seventy-three NNRs, 85 per cent of them in England. Of the total 28,350 ha, 4,450 ha were already held by conservation bodies. Shadowing NRIC's 'conservation areas', Huxley also proposed thirty-five 'scientific areas' covering almost 1 million ha. These did not accord so closely with NRIC's list since, even during the few intervening years, some sites had been badly damaged and were no longer worthy of recommendation. Of 174 sites listed by Rothschild in 1915,

Figure 4.1(a)   National Nature Reserves proposed by the Wild Life
Conservation Special Committee, 1947. *Source:* Sheail 1976

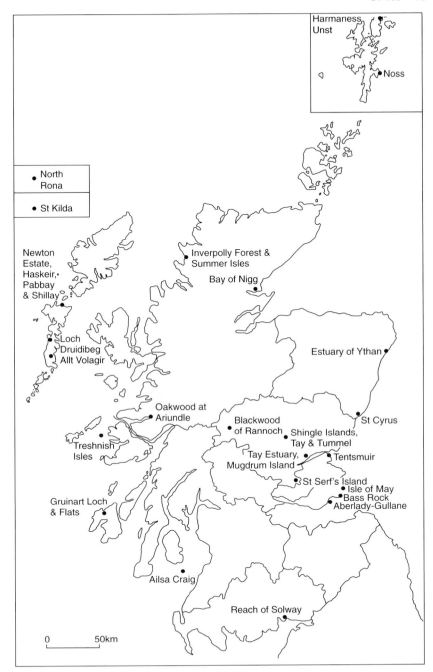

Figure 4.1(b) National Nature Reserves proposed by the Scottish Nature Reserves Committee, 1949. *Source:* Sheail 1976

twelve had been ruined: five by building development, three by afforestation, three by agriculture and one by the spread of bracken. 'Scientific areas' were expected to be sites in which, although not controlled by conservationists, nature conservation would be considered when any change in land use or any development was proposed. In effect, the vast majority of land in these areas was in the safe hands of agriculture and forestry and was expected to remain there for the foreseeable future. The possibility that such land uses could conflict with nature conservation was never entertained. They had always been at the heart of health and beauty in the countryside. But in the unlikely event of damage by farming or forestry, the machinery which enabled notification of land use changes would be sufficient safeguard. How disastrously that machinery would fail could never have been foretold. Indeed, even with the benefit of such numbing hindsight, we are still unable to redesign the machinery (p. 232).

Quite apart from the unexpected changes in land management techniques and the apparent weakness of the conservation argument against the economic one, a large part of the failure can be attributed to a lack of commitment by public and politicians alike. Huxley's recommendations were based upon the belief that the 'scientific areas' were as important as the NNRs. Likewise, his seventy-three proposed reserves were placed in no order of importance. It was a series of sites to give a balanced sample of organisms and habitat types, and each site played a vital part. Fail to designate one site and the whole system was devalued. Leave out one component and, like an aircraft, the whole thing would crash. Furthermore, Huxley had other ideas to broaden the foundation of nature conservation in England and Wales. He called for local authorities to be able to designate nature and educational reserves. He recommended the notification of 'Geological Monuments' and highlighted the need for a joint committee of conservationists and archaeologists to consider sites relevant to their interests. All of this would be in addition to the anticipated reserves within the National Parks.

The decision-makers were unwilling to countenance such a broad spectrum of new and potentially expensive ideas. So, although local authorities were given the power to designate local nature reserves (LNRs), they managed between them to set up seven, just seven, in the first fourteen years. There was no shortage of suitable educational reserves – a variety of committees had identified over five hundred by 1950 – but they fared no better than the LNRs. Geological sites had to battle for recognition under the umbrella of NNRs, and nothing at all was done to draw the archaeologists into the circle.

> The failure to include in subsequent legislation the National Park Reserves, Scientific Areas and Conservation Areas ... at once made the original list of proposed NNRs inadequate.
>
> (NCC 1977)

Huxley's vision, if not altogether shredded, was badly frayed before anything reached the statute books.

The findings of the Scottish nature reserves sub-committee, set up in January 1946 under James Ritchie, were similar to those of Huxley. *National Parks and the Conservation of Nature in Scotland* was published in 1947. The *Final Report on Nature Reserves in Scotland* did not appear until 1949, largely because there had been no earlier survey of potential sites from which to draw. Ritchie recommended four National Park Reserves as havens for wildlife, twenty-four NNRs for scientific study, and a series of LNRs for education and recreation. In addition, he listed twenty-two potential 'scientific areas' and called for the designation of 190,000 ha of Sutherland as a 'special conservation area' for its unique landforms and wildlife. Ritchie envisaged a Scottish Wildlife Service separate from any similar body in England and Wales. His report went much further than Huxley's in demanding direct government involvement in scientific research and the establishment of reserves.

But again, the Scottish situation was different. Ritchie made great play of the economic advantages to be drawn from nature conservation, but some naturalists found this hard to accept. The reserves they imagined were to be havens for wildlife and scientific researchers. Furthermore, they still had to counter some feeling, largely 'old hat' south of the border, that reserves would attract collectors and other unwanted visitors. They argued that nature conservation knew no political boundaries, fearing that a separate Scottish body would become the 'poor brother' while pressures in Scotland remained relatively slight. They won the day and the Nature Conservancy became a national body. Ironically, it was when pressures did increase north of the border, most notably in the flow country, that responsibilities for Scotland were devolved to a smaller, separate body (pp. 181, 211).

THE NATIONAL PARKS AND ACCESS TO THE COUNTRYSIDE ACT 1949

The Nature Conservancy (NC) predated the Act that established it by four months. In April 1948, Parliament announced plans to set up the world's first statutory, non-voluntary conservation body. It was born on 1 November 1948, with Cyril Diver as its Director and A.G. Tansley as Chairman. During these early months, the NC was responsible to the Agricultural Research Council. Today that seems an astonishing arrangement but again it makes the point that farming by good husbandry was considered to be closely akin to the conservation of nature.

The National Parks and Access to the Countryside Act of March 1949 confirmed the conservancy's official status, and a royal charter outlined its duties:

- to provide scientific advice on the conservation and control of the natural flora and fauna of Great Britain;
- to establish, maintain and manage nature reserves in Great Britain, including the maintenance of physical features of scientific interest;
- to organise and develop the research and scientific services related thereto.

It therefore had a dual function of research and conservation. The NC could own or lease designated NNRs or arrange nature reserve agreements with landowners and occupiers. Such agreements proved far more important than was expected, just as the overall number of NNRs far outstripped the recommendations of Huxley and Ritchie. The NC also had a power of compulsory purchase where a valuable area was sufficiently threatened and other means of safeguard had failed. In the event, the power was used only very occasionally when the relevant landowner could not be traced. Not until 1990 was it invoked for more sinister reasons (p. 170).

Another duty placed upon the NC by the Act was the notification to local planning authorities of Sites of Special Scientific Interest (SSSIs). These are areas of high-quality natural or semi-natural flora or fauna, or containing rare or endangered species or with key geological or physiographical features. Once scheduled, these sites receive special consideration when planning applications are made, and the local authority is duty bound to divulge details of such plans to the conservancy. As with NNRs, the aim was to form a series to include representative areas of each type; consequently, many sites of equal quality have never been designated. In respect of SSSIs, the conservancy's first task was to schedule those sites from the NRIC list of potential NNRs which failed to attract the latter status.

With the inception of NNRs and SSSIs under an official central agency responsible solely for nature conservation, and with local authorities empowered to establish LNRs, the stage was set for real progress on behalf of our wildlife. Although that progress was neither fast nor continuous, this aspect proved to be one of the strengths of the Act. For, in spite of its title, the National Parks and Access to the Countryside Act was largely a sham; 'a grievous disappointment to the amenity lobby' (MacEwen 1976). The aim was to designate National Parks in which access would be granted over wide areas of great natural beauty. Other regions, of similar landscape value, would be designated Areas of Outstanding Natural Beauty (AONBs). Although they would receive no special treatment and less emphasis would be given to public enjoyment, the label itself was expected to encourage recreational use. Within the parks, all public rights of way would be catalogued and publicised, and efforts made to open up further areas to general use. County Councils were given the power to provide access land either by order or agreement with the landowner or by outright purchase. This was likely to be pedestrian access only, and by-laws would protect occupiers' property. The inclusion of this clause was a direct result of the mass trespasses on the Derbyshire moors (p. 59).

The politicians made great play of the brave new world heralded by the Act. Lewis Silkin, who introduced the Bill, called it:

> a people's charter – a people's charter for the open air, for the hikers and the ramblers, for everyone who loves to get out into the open air and enjoy the countryside. Without it, they are fettered, deprived of their powers of access and facilities needed to make holidays enjoyable. With it, the countryside is theirs to preserve, to cherish, to enjoy and to make their own.

For the first time in history, the British people would be entitled to walk on other people's land, while landowners would be compensated for any damage caused. In the event, of course, the fanfares only served to heighten the disappointment of the Act's reality. None of our National Parks would qualify for the title by international standards, for they are all too intensively used to be of sufficient amenity value. AONB has proved to be nothing more than a paper designation (p. 227). In forty years, only 350 sq km of access land, 200 of them in the Peak District, were made available under this 'people's charter'. Most local authorities took the view that they had no open country of the type to which the Act referred! The level of commitment by official bodies to the new dream was dismal.

The problem was that, while the Act enabled the name 'National Park' to be applied, it failed to set up the machinery to administer a park on anything more than a parochial scale. Day-to-day running of a park rested with its National Park committee, two-thirds of whose members were appointed locally and one-third by the Secretary of State, advised by the National Parks Commission. Slightly different were the Lake District Special Planning Board and the Peak Park Joint Planning Board; autonomous and financially independent bodies which supervised planning applications in their respective regions.

It was clear from the start that any council worth its salt would consider local interests before national ones and, while they worked within the guidelines of the 1947 Planning Act, they were performing their elected duties. But the planning machinery had already begun to show its short-comings. Given the special characteristics of the parks, it was often to prove an impossible juggling act to balance local requirements for housing and employment with the National Park needs for recreation, amenity and quiet enjoyment. Various piecemeal efforts to improve the situation were to be tried in the coming years, but with little success.

To appease the doubters, the Act reconstituted the Standing Committee to become the National Parks Commission. This was the only national body to have an interest in the parks system as a whole. It set out to

Plate 4.2 To date, AONB designation has proved to be nothing more than a paper exercise, with land use practices developing here as in other 'unprotected' regions: Blackdown Hills AONB. Photograph by Peter Hamilton. Courtesy of Countryside Commission

represent all parties concerned with the preservation of natural beauty and the encouragement of recreation within the parks. It continues this work today, as the Council for National Parks, by lobbying the policy-makers, giving evidence at public inquiry and by seeking broader public appreciation of what the parks have to offer. But, throughout its history, it has been without teeth; a purely advisory body, frequently ignored in local planning discussions.

> The National Parks Commission was a political sop to a popular idea. The English national parks were really regional parks, and the commission had little money or executive power. Care of the parks and provision of recreational facilities have been quite inadequate. The men have been devoted but the governmental backing was timid and parsimonious.
>
> (Darling 1971)

So what real effects did the National Parks and Access to the Countryside Act have upon the British landscape and its wildlife?

5   1950s: Refuges and reconstruction

PLANNING FAILURE

By 1957, there were ten National Parks in England and Wales, totalling 13,618 sq km. For the next three decades, the movement appeared content, even though other potential parks, such as the South Downs and the Norfolk Broads, had been missed. Designation of AONBs was under way, but meant little in practice. In Scotland, Ramsay's five 'priority' parks became National Park Direction Areas (p. 69), extending to about 3,000 sq km but receiving little special attention. It was to be three decades before Regional Parks and National Scenic Areas would be devised in an effort to fill some of the gaps north of the border (p. 158).

The areas that were designated under the Act became popular and very well used. The dreams that were nurtured through the dark years of war unquestionably came true. Thousands of ordinary people were able to enjoy the clean air and open spaces, even though there was no complete freedom to roam. Car parks and picnic sites were filled, footpaths were well walked, information centres well attended. But all the parks were situated in the north of England or in Wales, none closer to London, for instance, than the Brecon Beacons. All consisted of mountain or moorland

Table 5.1   National Parks in England and Wales

Date	National Park	Area (sq km)
17.04.51	Peak District	1,404
09.05.51	Lake District	2,243
18.10.51	Snowdonia	2,189
30.10.51	Dartmoor	945
29.02.52	Pembrokeshire Coast	583
28.11.52	North York Moors	1,432
13.10.54	Yorkshire Dales	1,761
19.10.54	Exmoor	686
06.04.56	Northumberland	1,031
17.04.57	Brecon Beacons	1,344

Figure 5.1   National Parks (England and Wales) and National Park Direction Areas (Scotland)

habitat. Opportunities were missed, and those parks that did make the grade suffered from their own popularity. So heavy was the pressure in some areas that John Dower's fears became a costly reality; paths were eroded by trampling feet, parking space frequently outstripped, narrow lanes clogged. No longer could the visitor be sure of getting away from it all. It became increasingly like the proverbial honeypot, and maintenance costs soared. The problems were foreseen, but only by the far-sighted few.

Other problems arose in spite of and not because of the recreational nature of the parks. The most fundamental failure related to the new planning system – in overestimating its power to prevent damage to the landscape. Or it may be seen as an underestimation of farming, forestry and the mineral extraction industry. The Scott report assumed that, because there was to be a public hand in planning matters, the British countryside would remain forever in a state of equilibrium, based upon a system of farming which had that balanced state at heart. Nowhere was this happy condition less likely to falter than in the National Parks:

> in so far as the character of the country it is desired to include within a national park is determined by the type of farming ... it is essential for that form of utilisation to be continued with the proviso that in the case of a national park it becomes secondary to the main purpose which is public recreation.

Yet it was in the National Parks that the first indications were seen of an official lack of faith in the planning machinery. Landscape Areas Special Development Orders (LASDOs) were introduced as early as 1950. Relating then only to the Lake District, Peak District and Snowdonia, they required anyone, including farmers and foresters, to notify their intentions to erect *any* building. The authorities then had fourteen days to call for a full planning application if they feared that the design or materials were likely to be unsatisfactory. Immediately, the amenity lobby demanded the extension of LASDOs to all National Parks but it was to be thirty-six long years before this simple step was eventually taken (p. 225).

Quite apart from the world of farming, there were to be numerous developments which would highlight the total lack of real protection in the National Parks. Due to the nature of the parks, land is generally cheap, frequently rich in minerals and rarely populated to any great degree. In the economic world of the late twentieth century, this resulted in many clashes, many lost causes. Thus the National Parks landscapes carry military installations and reservoirs, power stations, quarries and major trunk roads. Such developments are anathema to the amenity lobby and the conservationist. They are nails in the coffin of the National Parks ideal, which is seen by many to have been cheapened over the years. MacEwen (1976) perhaps sums up a widely-held view: 'taken as a whole, the record of the parks in positive works and management before the 1974 reorganisation was pitiful'.

The problem revolved around the shortcomings of the planning machinery. Its missing links have been a gross disappointment to the millions who saw its introduction as a guarantee of perpetuity in our countryside.

> The idea of protecting country areas of special beauty has long since been accepted by the general public, yet planning control is generally neither firm enough nor clear enough to put it into practice: only continuous pressure from the public, e.g. from amenity societies, is likely to result in more effective control.
>
> (Gresswell 1971)

The trouble is that the planning machinery does not govern the things that matter most in the countryside. It concerns itself only with the specific proposal and not the overall picture; with the local rather than the national; with the immediate and not the long term. By and large, it relates to buildings, and only to some of those. Yet the visual effect, while important from an amenity point of view, is far less damaging to the health of the countryside than other consequences of bad development. This is what concerns the conservationist and, in the long run, the amenity lobby, too. The worst developments in the countryside since the 1940s have not related to buildings but to land use. The planning machinery does not govern these and yet:

> It is ... hard to see what essential difference there is between, say, planting up a stretch of heathland with conifers and building a small factory on it, or for that matter between felling an ancient wood and demolishing a listed historic building.
>
> (Mabey 1980)

The countryside is the work of nature, buildings are but the mark of man.

> conservationists of the 1930s and 40s saw farming as a buttress against landscape change; the main threats to the countryside were seen as ribbon housing development, factory buildings, mineral extraction and unsightly advertisement hoardings.
>
> (Shoard, in Lowenthal and Binney 1981)

So even farm buildings, let alone the techniques of land husbandry, escaped the attentions of the planners.

The changes in farming brought about by the war were not expected to last beyond it. The fourteen years following the Great War had seen the area of England under arable crops drop from over 2 million ha to 1.5 million ha. The consequent neglect of the farming landscape at this time had brought outraged complaints from the conservationists, who felt that farmers were neglecting their responsibilities. At the resumption of hostilities, the government intervened to encourage higher production. Grants were given to plough up ancient grassland, for example, and by

Plate 5.1 Eyesore buildings can always be removed, but the planning machinery remains powerless to tackle the real damage being wrought in the countryside. Photograph by the author

1942 arable hectarage had increased to 2.23 million. In March of that year, Ray Palmer wrote in *Countryside* magazine:

> The war has brought a great change over the face of agricultural England which cannot be without its effect on the wildlife of the countryside. Wiltshire was a great corn-growing county in the nineteenth century, but with the farming depression after the last war, the bulk of the arable land went down to grass. Now it is all coming up again, together with a good deal of downland never under the plough before. Over 120,000 acres [48,600 ha] of grassland have been ploughed up in this county since the autumn of 1939. The wildlife of the countryside is affected not so much by the Ministry of Agriculture ploughing programme as indirectly by the increased tidiness and improved drainage which is a concomitant of good arable farming.

But conservationist concern was not great. There was no reason to suppose that the rural landscape would not revert to its former state when peace came. A boom in agriculture could only result in a generally healthier countryside. If the meadows and the ponds were suffering, it was all in a good and, after all, temporary cause. So agriculture was allowed to develop separately from amenity and nature conservation, along the lines imposed by its own 1947 Agriculture Act. Nobody, perhaps, foresaw the changes that the Act would bring. Not even the farmers could have dared hope for the guaranteed prices and security of tenure which it offered in exchange for a degree of promised efficiency on their part. Modern farming came to Britain in 1947 and began its march across the countryside.

The conservationists gained more than the amenity lobby from the 1949 Act. Our series of NNRs, widely appreciated as field laboratories and havens for wildlife, far exceeds, both in number and area, what Huxley and Ritchie proposed. SSSIs too, have outgrown expectations. But the conservationists also stood to lose more from the effects of the Agriculture Act. For the new-born baby of the planners was far too weak to control the new direction of farming. In the ensuing thirty-five years, we would lose half our ancient deciduous woodland (more than was lost in the previous four centuries), half our lowland fens and mires, 60 per cent of our lowland heaths, 80 per cent of our chalk downs, 95 per cent of our hay meadows and 200,000 km of hedgerow (NCC 1984).

While NNRs could be purchased, leased or managed by agreement with the landowner, SSSIs fared worst of all. Some of the most valuable parts of our heritage, their label gave them no protection whatsoever. Farming and forestry have damaged a great many sites. Between 1950 and 1975, while many new SSSIs were being scheduled, 113 lost their status and eighty-seven were reduced in size as a result of damage. By the late 1970s, NCC estimated that 4 per cent of all sites were being lost every year through farming alone. This complete impotence of status forced a reappraisal of the situation in 1981 (p. 168).

The planning mechanisms were not designed to preserve our country-side from changes in working practices on the land. But why did the conservation machinery of 1949 not succeed more convincingly? There may be more SSSIs than expected but they suffered along with the rest of the countryside. We may have more NNRs than could have been hoped for, but many leases and management agreements are running out. Where is the protection in that? One reason was that many of the proposals for the Act were never instigated. Another cause was the tight rein upon which every government has kept the Nature Conservancy and its successors: the NCC and now the separate country bodies.

> To some people, the most serious and disappointing shortcoming in nature conservation practice has been the failure to translate into reality the broad and integrated concept of conservation which was the great vision.
>
> (NCC 1984)

Thus, there were no 'conservation areas', no National Park Reserves or Local Educational Reserves. Very little, in fact, was achieved for the wider countryside in general. This has been attributed to the powerful vested interests and influential contacts of the landowners and their resentment of an infant conservancy presuming to advise them on management techniques. Their paths of progress were inherently anti-conservationist and they were shown very little economic reason to preserve wildlife. The weakness of the conservancy was no small factor in this. It battled on with mixed fortunes under the vagaries of the political climate until it was finally dismembered in 1991. Throughout its life, it suffered from under-funding and lack of influence.

> Nature conservation has in the past sometimes conducted its business on too apologetic and timid a note. Such a tendency to submissive posture is a recipe for retaining a low peck-order position in the league of land and resource use interests.
>
> (NCC 1984)

For the first forty years of its existence, the official body for British conservation depended, as did the entire movement, on the efforts of a few determined individuals.

> My experience while working for the Conservancy showed me that nearly all the effective actions of that body were due to determined efforts of individual people who identified themselves with particular problems and promoted them until they had achieved their objectives. Directorates and committees can destroy initiatives and they can give helpful support, but in the last resort it is only the commitment of individual enthusiasts which produces the goods.
>
> (Moore 1987)

Thanks to the foresight of private conservationists, and precious little thanks to career civil servants or ministers, Britain has a tried and established piece of machinery to funnel ecological advice into policy decisions. . . . in this field, Britain did have a head start over the rest of the world.

(Fitter and Scott 1978)

POST-WAR RECONSTRUCTION

The war years had interrupted the momentum of the voluntary conservation movement. Membership figures levelled out and the number of new initiatives dropped as Britain understandably turned its attention to more pressing matters. But, with the return to peace, the time was right to renew the thrust. Conservation fitted into the ideal of a brave new world very comfortably. Britain had pulled through again and her people deserved to receive their promised land in good heart. Public access in search of a new knowledge of the countryside would rely upon its conservation and careful management. With the 1949 Act and the establishment of the Nature Conservancy, official recognition of the cause had been granted. The voluntary movement, spurred on by this seal of approval, now needed to break new ground in its campaign on behalf of our wildlife. For ten years following the war, there was much shuffling and regrouping within existing ranks and an infusion of much new blood. The movement was preparing itself for a concerted effort on an ever-widening front.

The concept of an International Union for the Protection of Nature dated back to 1913, when seventeen countries had signed a 'diplomatic instrument' proposed by the Swiss protectionist, Paul Sarasin. His experiment had failed under the vagaries of international relations. The Second World War, though, diminished the size of the planet in many eyes and, combined with the moves towards a new start, this brought European conservationists together. No longer could they rely upon isolated efforts in individual countries. Nature knew no national boundaries. Furthermore, they realised that human civilisation depended upon the natural environment; that natural beauty was vital to spiritual fulfilment; that ever more effective means of exploiting the world's resources were about to exceed their ready availability and that living standards would inevitably suffer; that education could turn the trend to one of careful, conservation-minded use; and that there was a pressing need for a 'single effective international agency' concerned with the protection of nature.

Within a year of armistice, members of the Huxley Committee attended exploratory meetings on the continent and, on 5 October 1948, the International Union for the Protection of Nature (IUPN) was officially founded. Its objects were to:

- encourage and facilitate international co-operation between governments and interested bodies in the 'Protection of Nature';
- promote national and international action on;

  (a) the preservation of wildlife and the natural environment, especially species threatened with extinction,
  (b) the spread of public knowledge about the 'Protection of Nature',
  (c) the promotion of an education programme on the 'Protection of Nature',
  (d) the preparation of international draft agreements and a world-wide convention for the 'Protection of Nature',
  (e) scientific research on the 'Protection of Nature';

- collect, analyse, interpret and disseminate information about the 'Protection of Nature'.

The Union was 'the first durable expression of the popular concern for the environment on a worldwide basis' (Durrell 1986) and left the stage set 'for the world-wide spread of environmental and wildlife conservationist attitudes' (Fitter and Scott 1978). The Constitutive Act was signed by eighteen governments, seven international organisations and 107 national conservation groups. As governments, official bodies and voluntary groups could enrol, there existed great potential for a valuable forum to direct the enthusiasm and expertise of conservationists the world over to particular problems wherever they were faced. Until this time, many individuals with conservationist ideas had been battling alone or privately nursing their visions without so much as a national body to support their broad aims. But initial progress was slow. Added to the inevitable teething problems of an infant organisation working on a world-wide scale, a severe lack of finance made practical development impossible. Like the NC in Britain, its early survival depended wholly upon the efforts of individuals. In many countries, official apathy was apparent and Britain was one of the nations shamefully absent from the list of governments supporting the Union at its inception.

> I virtually went down on my knees to Cyril Diver (then first director of the newborn Conservancy) – whom I greatly admired – to try and persuade him to change his mind, but he stubbornly maintained that he was only concerned with the United Kingdom and that his interest in conservation ended at the Channel.
>
> (Rothschild 1979)

In retrospect, that seems a strange attitude for the official conservation body of a country which was supposed to be in the forefront of the movement. But the conservancy was suffering teething troubles of its own. In truth, things were progressing too swiftly for the British conservation movement and its government.

In spite of such intransigence, the Union did gain support and, in 1956, became truly international with the addition of North America to its list of signatories. At this time, to better reflect its concern for the whole environment, it became the International Union for Conservation of Nature and Natural Resources (IUCN). After the WWF was set up in 1961 to finance its activities (p. 114), IUCN developed rapidly on a global scale. By 1985, it had 541 members in 116 countries, comprising fifty-eight states, 124 government agencies, 349 non-governmental organisations and ten affiliates. Since 1983, it has run a major data bank on endangered species and protected areas at its Conservation Monitoring Centre in Cambridge, England. In 1988, it changed its name to IUCN – The World Conservation Union.

The SPNR was rekindled after its years in the doldrums. Although it remained committed to the idea of reserves, it could afford to press in other directions now that a series of NNRs was to be established by a government body. Its focus shifted to the growing band of local trusts, which lacked any central advisory body (p. 52). In 1957, the decision was made to become the umbrella organisation for these trusts. Not only could advice be given but also some financial support in the purchase and management of reserves. SPNR had become increasingly embarrassed by an ever-growing surplus of funds derived from bequests through its doldrum years. Its offers to assist in the purchase of reserves had rarely been taken up. So, in 1958, the County Naturalists' Trust Committee was formed to encourage the expansion of the local movement both geographically and further into the arena of reserve purchase and maintenance.

In 1947, the Botanical Exchange Club (p. 37) regrouped and named itself the Botanical Society of the British Isles (BSBI). With the increasing watchfulness of the local trusts, it was able to spend less time on individual sites of importance and to concentrate more deeply on matters of wider concern. It formed a Threats Committee (later renamed the Conservation Committee) to highlight to the official bodies the growing risks to our flora, unprotected as it was by any real legal provisions.

On 10 November 1946, a young man named Peter Scott, whose only claim to fame was that he was the son of the Polar explorer, opened his Severn Wildfowl Trust on 7 ha of wetland at Slimbridge in Gloucestershire. This was to be an observation and research centre for the intimate study both of captive wildfowl within the collection and of wintering wildfowl on the Severn Estuary. From the outset, the Wildfowl Trust has succeeded in bringing people and wildfowl together, to the inevitable advantage of the conservation cause. Its success in education has equalled, if not exceeded, its enviable research record. Furthermore, much good work has been done in practical conservation. New wetland habitats have been developed – the trust refers to its Martin Mere reserve as having been 'reclaimed from agriculture' – and endangered species have been bred in captivity for return to the wild in various parts of the world. 'The

Plate 5.2 The Wildfowl and Wetlands Trust has become a leader in research and international co-operation to save this declining habitat, but has never forgotten the value of education and grass-roots support. Photograph by the author

Wildfowl Trust was the first and most successful of the postwar organisations in Britain devoted to the conservation of individual species or groups of species . . . . Similar trusts were later founded to help conserve pheasants, hawks and others' (Fitter and Scott 1978). In 1989, it adopted a new name, the Wildfowl and Wetlands Trust, to reflect its broader concerns. It now has eight centres in all parts of the UK which receive up to 750,000 visitors a year. Membership stands at 70,000 and Sir Peter Scott has become a household name in his own right for his work in conservation.

Shortly after the trust was formed, ICBP set up an International Wildfowl Research Institute as an information unit and research body. It subsequently became known as the International Waterfowl and Wetland Research Bureau (IWRB), now Wetlands International. Since the early 1960s, it has run global conferences on the study and conservation of wildfowl, the most notable being that at Ramsar in 1971 (p. 139).

The new blood with which the movement was being infused also included organisations with an interest in our lesser known wildlife. Less attractive in so many ways than plants or birds, our reptiles and amphibians found their champions with the formation of the British Herpetological Society (BHS) in 1947. Resident in mainland Britain are two species of toad, three of frog (two of them introduced) and three each of newt, snake and lizard. In Ireland, only the common lizard and the smooth newt are native, while the common frog was introduced in 1699. All are highly susceptible to loss of habitat, disturbance, heathland fires and collecting. The society carries out research and field study into these threats and the best ways to counteract them. A programme of captive breeding provides stock for study, for supply to collections and for reintroduction to the wild. An education programme seeks to improve the image and treatment of these creatures by all sections of the population, from politician to school-pupil.

On behalf of our larger species, the Mammal Society was established in 1954. It encourages the amateur and professional study of mammals worldwide, but with particular reference to British species, and seeks to increase general appreciation of them. Important work has been carried out on behalf of deer, otters, bats, whales and dolphins.

With its increasing size and variety, the conservation movement began to feel the need for a public relations body to encourage general interest in the cause and to put its case to the policy-makers. Individually, the existing societies could not afford a concerted publicity campaign. Many lacked the expertise and their interests were generally too narrow. The NC could not involve itself in political debate. The SPNR was busy marshalling the county trusts. So, in July 1958, the Council for Nature was formed. Most national and local societies quickly lent their support. Within four years, 292 groups had affiliated, representing over 80,000 individual members. The reviewed constitution of 1968 outlines the council's main aims throughout its existence:

> the function of the Council shall be to co-ordinate the views and information of the voluntary bodies in the United Kingdom concerned with the conservation of nature and the study of natural history; on behalf of these bodies collectively, to make representations to Her Majesty's Government and other authorities whenever necessary and to act in a central consultative capacity as required; both on particular questions and in general, to keep the need for nature conservation constantly before the public; and increasingly to interest young people in safeguarding the country's natural beauty and wildlife.

The Council for Nature scored many successes. It brought popular pressure to bear upon such activities as the gassing of badgers, the culling of seals and the use of pesticides – causes that were certain to promote general concern. National Nature Week proved a catalyst for a myriad of new ventures (p. 114). But financial constraints always hampered the council, even after 1976 when it reduced its staff and farmed out many of its activities to other bodies. Furthermore, the set-up became unwieldy; some conservation bodies were not represented, while many affiliated groups had little to do with nature conservation. The Council for Nature ceased to function in 1980 and its affairs were taken on by CoEnCo (p. 122) and the newly formed Wildlife Link (p. 165).

The most successful product of the council's many worthwhile endeavours was the British Trust for Conservation Volunteers (BTCV). Set up in January 1959 as the Conservation Corps, the aim then was to involve young volunteers in practical management work on NNRs. From its first field day on Box Hill in Surrey, the corps' activities expanded rapidly. Much of the early work was carried out from residential camps in remote areas but day-long or weekend tasks within 80 km of London became increasingly popular. By 1970, when BTCV took over the running of the corps, 63,000 work-days had been performed on over 1,300 tasks.

The success of BTCV can be attributed to two factors. The first is the growing inclination of individuals to perform voluntary work (p. 255). The second is that its development has run parallel to the deepening concern for our environment. For, while much of the movement in the 1950s was turning increasingly towards the scientific, the aesthetic appeal of nature continued to hold it in good stead. For all the survey and study, the counting and comparison and the compilation of records, nature's popular image was intensifying as never before. The media, especially television, was part cause and part effect of this. Newspapers have generally relegated their 'Country Corners' to the bottom of an inside page. General interest magazines have done better, particularly in recent years, but purely wildlife periodicals have not been successful over the years. Radio has much to offer but the attraction of the natural world depends upon more than sound and word. With the addition of sight, albeit secondhand, television has consistently proved the most successful medium. Although the experience is insular and detached, much can be made of the intimate insight that the camera provides: a close-up of a honey-bee, life inside a nest box, the accelerated bursting of a flower or the retarded flight of an owl, an underwater world or a distant desert.

The conservationists were quick to realise the potential of films and television. James Fisher, Ludwig Koch, Peter Scott and others appeared on BBC in its first three years, 1936–9. The RSPB had its first film made in 1950 and, two years later, formed its own film unit in the light of that success. By 1984, 214 million people world-wide were thought to have watched its films. Likewise, the television producers understood the capacity of the natural world to enthral an audience. On 14 June 1955 the first programme in the *Look* series was transmitted. So popular did the series prove that the BBC Natural History Unit was formed as a direct result in July 1957. Within a year, a number of nature programmes were being given a regular slot: *Faraway Look*, *Out of Doors*, *On Safari*, *Diving to Adventure*.

In 1959, the BBC awarded a grant of £25,000 over five years to the Council for Nature to support the setting up of an Intelligence Unit. The Unit, in line with the Council's aims, was to publicise the cause of conservation but, specifically, the BBC wanted to encourage the making of natural history films. It was a practical step by the media to bring the two interests together to their mutual advantage. The natural world has maintained a high position of importance to television ever since and continues to attract unprecedented numbers of viewers. First-class commentators have played their part. David Attenborough, through series such as *Life on Earth* and *The Living Planet*, has awakened great appreciation of our natural riches. Until the 1970s, it was sufficient for television purely to document the spectacular, but that decade ushered in a real concern for its conservation. The media, especially television, acted accordingly and began to cover the theory and practice of nature conservation. As a result, a 1983 opinion poll showed that 68 per cent of Britons relied upon television for information on wildlife and environmental problems. Newspapers were used by 43 per cent as a source of intelligence while only 11 per cent listened to radio and 10 per cent read magazines on conservation issues. Educational establishments, specialist societies and pressure groups each polled a mere 2 per cent. There is no doubt that the natural world's aesthetic appeal, viewed through the eye of television, has done much to assist its conservation.

Some important initiatives were being taken in the publishing world. Early in the century, Warne had begun to produce its excellent *Wayside and Woodland* series, which ranged from birds and trees to spiders and pyralid moths, and which still warrants inspection today. But it was Collins's highly acclaimed *New Naturalist* series, that really brought the natural world within the grasp of the layman. In twenty years, it grew to over fifty volumes which covered all aspects of natural history and, in due course, came to have its own history written (Marren 1995). The series played a part in guaranteeing a popular market for the equally successful field and pocket guides which also began to come off the Collins presses during the 1950s. These guides have been widely imitated but rarely bettered and they now extend to the flora and fauna of many parts of the globe. Oxford University Press produced a fine series of guides during the 1960s, many of which have since been reissued in pocket version. Field guides traditionally rely upon the work of accomplished illustrators but Pan, in the late 1970s, published a successful series of photographic guides, starting with Roger Phillips's *Wild Flowers of Britain*. What photographs lack in clarity, they make up for in realism, since animals and plants in the field are never seen frozen onto a white page. Yet there is a certain attractive charm in good artwork which undoubtedly sells books. In 1965, when the Rev. W. Keble Martin published his *Concise British Flora in Colour*, after sixty years preparing the plates, it was immediately acclaimed a masterpiece and has remained a leading text to this day.

REFUGES AND PROTECTION ACTS

By gaining official acceptance under the 1949 Act, nature reserves secured their place in the future of British nature conservation. The first NNR was declared on 1 November 1951. Beinn Eighe in Ross and Cromarty is a magnificent mountainous area of relict Caledonian pine forest with deer, pine marten, wild cat and fascinating flora. Since only 8,000 ha of the original 1.2 million ha Caledon Great Wood remains, the value of the reserve is beyond question. Yarner Wood in Devon and Moor House in the Lake District National Park soon followed as the first NNRs in England. By the end of 1952, there were nine NNRs covering 8,900 ha and, by the end of the decade, the tally was eighty, totalling 55,765 ha – a fair reflection of the alacrity with which the newborn conservancy set about selecting its reserves.

Unfortunately, local authorities did not follow the lead. In 1952, the first LNRs were declared; in Scotland, at Aberlady Bay, East Lothian; in England, at Gibraltar Point, Lincolnshire. But momentum was lacking. By 1975, there were two LNRs in Scotland; two in Wales and thirty-six in England. These forty covered only 5,589 ha, just 5 per cent of the area of NNRs at that time.

The FC, to its credit, fell in with the times by declaring Forest Nature Reserves in addition to its existing Forest Parks. The first, announced in 1954, consisted of 80 ha at Waterperry in Oxfordshire, to be administered jointly with the NC in the interests of timber production, nature conservation and ecological research.

In addition to these various categories of reserve, a new thrust was made towards national and regional wildfowl refuges:

- to serve as a strongpoint for certain species to ensure their status in an area;
- to form wildfowl reservoirs;
- to be used to mitigate local effects of overshooting.

Sanctuary Orders had already established a handful of refuges: on the Exe Estuary and at Tamar Lake in Devon, at Walmsley in Cornwall and Radipole Lake in Dorset. But heated arguments between the wildfowlers and the conservationists during the run-up to the 1954 Protection of Birds Act brought their differences of opinion to a head. 'The discussion in Parliament and in the press at the time of this Act revealed a horrifying state of ignorance, confusion, suspicion and emotion . . . ' (Nicholson, 1961, quoted by Harrison 1973). Ignorance and confusion reigned because of the total lack of any meaningful figures on the effects of the sportsman on wildfowl populations; suspicion and emotion stemmed from the divergent attitudes of the two parties towards our wildfowl.

But both sides had the same interest at heart: the preservation of a healthy stock of 'quarry' whether to watch or to shoot. Thus it was that when the two sides came together, much headway was made. They came together as the Wildfowl Conservation Committee, with representatives from WAGBI, the Wildfowl Trust and the NC. This was the committee given the task of drawing up the series of refuges. Its unanimous agreement in January 1955 to designate the Humber Estuary as the first in this new generation of refuges was a resounding vindication of open discussion and restrained ideals and a grievous disappointment to a public and press which had been alerted by the earlier antagonism. Subsequent Sanctuary Orders were made at Southport in Lancashire, Wicken Fen in Cambridgeshire and Hamilton Low Parks, Lanarkshire. Other refuges came to be protected by NNR and LNR designation.

The Wildfowl Conservation Committee went on to show just how much can be gained by bringing heads together to reason rather than to clash. Inspired by the discovery of common ground, WAGBI and the Wildfowl Trust jointly undertook a seven-year project on the food preferences of British ducks, establishing, in 1956, the Sevenoaks Wildfowl Reserve in Kent to put their researched theory into practice. Food plants and cover vegetation were planted and safe nesting sites provided in the form of floating rafts, and the birds began to arrive in increasing number and variety.

The winter monthly national wildfowl counts, begun in 1949, grew and extended to the continent, so providing a firmer base for discussion than the aesthetic one. In the aftermath of the conflict over the 1954 Act, WAGBI began to breed certain species of duck for release into the wild and followed this by promoting a series of local reserves by private agreement, entirely separate from any already in existence. By the early 1970s, over eighty such reserves covered 4,000 ha.

From 1954 onwards, the wildfowlers worked hard, in the interests of both their image and their sport, to nurture a more conservationist attitude. They succeeded, to the inevitable advantage of our wildfowl, and left the conservationists struggling to move further onto this broad area of common ground.

Wildfowlers may well be forgiven for holding reservations during the debating stages of the Protection of Birds Act. It was promoted by Lady Tweedsmuir whose Private Member's Bill remains the longest on record. Supported by many influential people, it was, when it became effective on 1 December 1954, a very powerful piece of legislation for its time. It repealed no fewer than fifteen previous Acts and put an end to the need for county lists of protected species. At last, the hard-fought battle for a Black List system (p. 48) had been won. The Act began with the premise that all wild birds, their eggs and nests were protected and that any person commits an offence who wilfully:

- kills, injures or takes, or attempts to kill, injure or take any wild bird; or

- takes, damages or destroys the nest of any wild bird while that nest is in use; or
- takes or destroys an egg of any wild bird; or
- has in his possession or control any wild bird recently killed or taken unlawfully.

The maximum penalty at this time was £5 for each egg, nest, skin or bird, increased to £25 and the risk of a month's imprisonment in the case of rarer species listed on the First Schedule.

From this broad starting point, the exceptions were introduced, whereby authorised persons in certain circumstances could take birds, etc. on the Second Schedule (twenty species) or, outside certain close seasons, birds on the Third Schedule (twenty-two species). Thus the interests of landowners and sportsmen were preserved. The Act also made it an offence to sell, offer for sale or possess for sale any live wild bird on the Fourth Schedule, the eggs of all British breeding species and any dead wild bird or skin taken unlawfully. There were provisions for the Secretary of State to establish bird sanctuaries with the landowner's consent; controls on the import of certain birds and eggs; and a requirement to provide caged birds with sufficient space to stretch their wings. It is a sobering thought that, in the middle of the twentieth century, the Act was also outlawing 'springes, traps, gin, snare, hook and line, poisoned or stupefying bait, floating container holding explosives, net, baited board, bird lime', together with the use of live decoys and of artificial light for trapping!

Comprehensive protection for one type of wildlife had at last reached the statute books and Vesey-Fitzgerald (1969) is perhaps mistaken to suggest that 'these Acts, based on situations and events now long past, are not really relevant to the present'. The 1954 Act was the natural progression from earlier legal moves, and the bird protectionists still light the way. The situations and events may have changed but human compassion and sensibilities have not, and it is upon these that the conservation movement has always relied.

> For the moment, cool rationality is the fashionable style; and the voice we can hear behind the major pieces of wildlife legislation in the seventies ... is more dispassionate and, I believe, ultimately less powerful than that which energized the bird protection Acts 100 years ago and through them helped to begin the whole organized conservation movement.
>
> (Mabey 1980)

So sure was the foundation of the conservationist ethic that the 1950s began to see a noteworthy trend in more general legislation reflecting a shift in the attitudes of the law-makers. The conservation movement must take credit for instigating the wind of change. Many Acts of Parliament began to carry a clause such as:

the Board or the Minister ... having regard to the desirability of preserving natural beauty, of conserving flora, fauna and geological or physiographical features of special interest ... shall take into account any effect which the proposals would have on the natural beauty of the countryside or on any such flora, fauna, features ...

The Electricity Act of 1957 and the Opencast Coal Act of 1958 are examples from a number of statutes which now placed a duty upon the powers-that-be to consider the effects of their work upon our landscape and wildlife. Efforts to comply were made but not as a high priority, and natural resources continued to suffer heavily at the hands of industry. In the same year as the Electricity Act, the Central Electricity Generating Board, against fierce opposition from the NC and the RSPB, won approval for a nuclear power station on the NNR and SSSI of Dungeness in Kent. Education to alter basic values was a slow and painful process. Only in the 1990s, after coercion from the European Community, did Britain eventually begin calling for Environmental Impact Assessments of major industrial developments (p. 195).

POLLUTION AND PESTICIDES

In the early 1950s, much concern from conservationists but not, apparently, from industrialists revolved around the pollution of our seas, rivers and atmosphere. It was estimated that oil pollution was responsible for between 50,000 and a quarter of a million seabird deaths annually around British coasts. The ICBP alerted other concerned interests and together they formed an Advisory Committee on Oil Pollution of the Sea. It was through the efforts of this Committee that the International Convention for the Prevention of the Pollution of the Sea by Oil was set up in 1954. But the Convention was as unwieldy as its name and proved to be largely ineffective. The 1955 Oil in Navigable Waters Act was likewise futile. Oil pollution at sea increased through the 1950s and 1960s, taking devastating toll on marine life.

River pollution too was a disgrace. A 72 km stretch of the Thames was entirely devoid of oxygen. Just about all river life had abandoned it, and the natural breakdown of waste had ground to a halt. Many other rivers in and below industrial areas wore the same sad face of suds and foam; factory waste washed directly into them with no prior treatment. The Rivers (Prevention of Pollution) Act 1951 set up River Purification Boards charged with ensuring the quality of water resources and with monitoring and controlling pollution. Any new effluent discharges to non-tidal waters had to be approved. In 1960, the requirement was extended to discharges into tidal waters and, a year later, to those that had started before 1951. In that way, limits and conditions could be attached, but no financial charge was made and, most ominously, details of discharges were not made public. Over the years, the boards did encourage an improvement in water quality but progress was not always brisk. Thus the voluntary bodies found themselves concerned with the problem. The Anglers' Co-operative Association, for example, was set up to fight for financial recompense through the civil courts for angling clubs and individuals whose waters were polluted by industry. The association scored some notable triumphs, making legal history in a successful case against three major polluters of the Trent. Many offending factories cured their problems before any legal action was taken against them. When the Water Resources Act 1963 was passed, it placed upon the river authorities a duty to consider the preservation of natural beauty and the conservation of flora and fauna.

But the overriding issue of the early 1950s was air pollution. Much of it was caused, not by industry, but by domestic fires. Various societies had sprung up over the years and eventually merged to become the National Smoke Abatement Society, now the National Society for Clean Air and Environmental Protection. Manchester became 'smokeless' in 1946 and Coventry in 1951 but, as a rule, the problem was allowed to continue out of hand. It took the infamous London smog of December 1952, which claimed 4,000 human lives, to jolt the authorities from their indifference. The Beaver Committee on Air Pollution was convened the following year and the outcome was the Clean Air Act of 1956. This, the first legal provision in the world to control domestic smoke – it also regulated industrial emissions – succeeded in reducing smoke levels in London by over 75 per cent in the next two decades. Even so, many local authorities were tardy in its implementation, and 750 people in London alone died through the effects of smog in 1962. The toll of such pollution on our wildlife will never be known, if only because the consequences manifest themselves in so many ways. Sunlight is reduced, affecting plant growth and animal health. Some pollutants upset plant metabolism; earlier this century, it was found to be impossible to cultivate coniferous trees on the polluted uplands between Sheffield and Manchester. Air pollution can weaken animal bone by increasing the uptake of fluoride. Most devastating is the acidification of soil and water, a matter of particular concern in the closing years of the century (p. 186).

Pesticides were the other pollutants viewed with severe apprehension in the 1950s. Unlike smoke and sewage, they were relatively new on the scene. Their effects were less immediately evident, more sinister and more damaging to our wildlife. Nor were they really understood; not by the scientists or the manufacturers, and least of all by those men who sprayed them on the fields. Pesticides in general enjoyed a fine reputation. DDT had saved millions of human lives worldwide and was seen as 'the greatest boon that mankind has ever had' (Mellanby in Chisholm 1972) and 'a paragon of all the virtues' (Moriarty 1975). Other substances controlled or eradicated crop-pests and weeds, saving hours of backbreaking work and increasing farm yields to impressive levels. So no suspicion fell upon these substances when things began to go awry in the countryside.

It was the number of bird deaths that sounded the first warnings. Insects were expected to die, so too were many plants. Mammals were less obvious to the human eye, and so the birds' high-profile popularity, which had brought them early protection, now signalled the dangers of pesticide use. In the early 1950s, dinitro and organophosphorus compounds such as Parathion were killing large numbers of farmland birds. Sprayed to combat aphids on vegetable crops, they were extremely poisonous both to people and to other species. They were not, however, as persistent as some other chemicals. In 1955, a government research report, *Precautionary Measures against Toxic Chemicals used in Agriculture*, called for stricter controls upon these substances in the light of the damage they were causing.

In August of that year, the Ministry of Transport decided to limit the spraying of chemicals on verges to major roads only. The farmers, too, were willing to help but, as a result, the organochlorine compounds began to gain favour. These included aldrin, dieldrin, endrin, heptachlor, DDD and DDT. But these compounds were highly persistent and only broke down into harmless remnants over very long periods of time. Consequently, the presence of these poisons in the countryside reached tremendous proportions. Through the food chain, they entered the bodies of animals, including man. Predators near to the top of the food chain, while not consuming the accumulated quantities in their prey, as is commonly believed, nevertheless began to succumb very rapidly to the effects of these compounds. The otter hunts began to notice drastic declines in their quarry. Post-mortems on badgers showed increasing levels of pesticide residue. In the last winter of the decade, 1,300 foxes died in East Anglia as a result of pesticide poisoning. Numbers of birds of prey fell drastically. Kestrel numbers plummeted. The sparrowhawk, which had so ably survived all the unhealthy attentions of the gamekeepers over the years, was afforded legal protection for the first time in 1963 because it had become so uncommon. Peregrine populations dropped to below half their 1939 levels. In fact, all creatures were affected; all wild populations suffered.

Yet only slowly came the dawning realisation that chemical poisoning was the cause. A Pesticides Safety Precautions Scheme existed to ensure that only 'safe' compounds would be freely available. Apparently, it was not succeeding. So the government charged the NC to look into some of the questions surrounding pesticides. The Toxic Chemicals and Wild Life Group was set up in January 1960 to investigate such matters as the long-term effects of sub-lethal poisoning, declines in animal populations and the extent to which populations of different organisms are able to adapt to different chemicals. The group performed some valuable and enlightening work but not without difficulty. The scientific research on reactions caused by certain chemicals was straightforward enough. Alarmingly, what was lacking was any real account of our common wildlife; for instance, the wildlife communities of the hedgerow. To add to this, when the group began to investigate our hedgerows, they found them not to be as widespread as was commonly thought. They were disappearing at a very fast rate. So we see the value of basic survey and census work on our most common forms of wildlife and habitat even when they seem to be unthreatened.

When Derek Ratcliffe began to survey breeding peregrines in 1960, he too uncovered a story of decreasing numbers which took just about everybody by surprise. Ironically, the survey was sparked by complaints from pigeon fanciers that increasing peregrine numbers were adversely affecting their sport. It is no wonder that the effects of pesticides were being underrated, when a drastic fall in the numbers of our most tantalising bird of prey could go almost entirely unnoticed.

Ratcliffe found, following ten years of population increase after the war, that peregrine numbers were falling in 1955, most markedly in southern England. Moreover, birds attempting to breed were managing a success rate of only 26 per cent. Aware of the growing conflict surrounding organochlorine pesticides, Ratcliffe soon suspected, and found, a correlation between their use and this latest population demise. Just how the two married up was not immediately clear. The popular story is that Ratcliffe was watching a bird settle on her nest when one of the eggs cracked beneath her weight. Certainly, figures were showing that the number of clutches containing broken eggs had reached 39 per cent, a tenfold increase over the records of the 1940s. Closer analysis of the eggs showed them now to be thinner-shelled than previously. Further investigation of the eggs of seventeen species of bird highlighted definite thinning in nine of them: peregrine, sparrowhawk, merlin, shag, golden eagle, hobby, rook, kestrel and carrion crow – six birds of prey, two crows and the fish-eating shag. The failure of the eggs together with the outright mortality through high chemical concentrations in the body combined to throw some populations into steep decline.

As the decade closed, so too did the golden years of the pesticide industry. High praise was replaced by grave reservation; unquestioning encouragement by suspicion. While chemical control still had much to offer, tighter restraints and closer supervision became the norm. One of the gravest threats to British wildlife had ended its free flight of the 1950s and would now have to come to heel. We must never underestimate the sterling work of the scientists and conservationists in alerting themselves and convincing others of a sinister threat so new that an ignorance of it could have been forgiveable (see Sheail 1985). The work of Derek Ratcliffe and the NC on organochlorine pesticides is 'arguably the most outstanding scientific contribution yet to nature conservation in Britain' (Adams 1986).

6 1960s: New conservationists and the Countryside Acts

CHEMICALS IN THE COUNTRYSIDE

The confusion over pesticides spilled into the 1960s. Sufficient clamour was generated by the conservation bodies, and sufficient concern voiced by public opinion and some politicians, to bring about a voluntary ban on the use of aldrin and dieldrin in 1962. The same year saw the launch in America of Rachel Carson's shattering book, *Silent Spring*, which was published in Britain in 1963. The book was a frightening dossier on the use, misuse and deleterious effects of pesticides, largely drawn from American examples. It was written, of course, with great bias, but the chemical industry mounted a powerful and vitriolic counter-attack, holding the bounteous benefits of their products like a banner at the helm, and so all unfair advantages were effectively cancelled out! Nicholson (1970) sees the book as 'probably the greatest and most effective single contribution hitherto towards informing public opinion on the true nature and significance of ecology'.

What Carson succeeded in doing was to stir up a public outcry for the truth. The voluntary sector was already researching the problem. The RSPB and BTO organised a Joint Committee on Toxic Chemicals which worked closely with the Game Research Association. The pesticide producers, too, appeared to be concerned at the allegations levelled against them. In March 1964, they got together with the conservationists at a symposium on 'Agricultural Chemicals – Progress In Safe Use'. The NC was engaged in its research but the public demanded more from the government. The Cook Committee was formed. It reported its findings in 1964 and, within a year, the ban on aldrin and dieldrin became compulsory – but for some uses only. For remaining uses, the ban continued as a voluntary measure.

This blatant contradiction typifies the whole question of pesticides during the 1960s and beyond. Nagging reservations were held about the power of these compounds, yet bans were partial, piecemeal and voluntary. Vesey-Fitzgerald, a well-respected naturalist, wrote of the 1965 ban: 'This was a magnificent example of public spirit on the part of the farmers'. Written in 1969, was this sincerely meant or was it an attempt at building bridges between the two sides in the hope of further concessions? In that year, the Pesticides Advisory Committee was demanding a total ban on all organo-chlorine compounds. Research was showing, for instance, that the livers of badgers contained up to 100 times the 'normal' concentration of dieldrin. The worst-affected individuals were seen gnashing their jaws, squealing loudly and acting as if drunk, as their muscle co-ordination failed. Death came in convulsions. Was the ban so very public-spirited in the light of this not irregular occurrence in the British countryside? Three months after these findings were made public, the government renewed the voluntary ban. While the manufacturers had withdrawn certain compounds from their lists, others of equal potency remained in widespread use. Yet Darling, in the 1969 Reith Lectures, said: 'Britain is showing exemplary action over curtailing use of the organochlorine pesticides'.

Writing in Goldsmith and Hildyard in 1986, Rose stated that 'Britain's environmental record in controlling pesticides has been nothing less than deplorable'. But Rose, unlike the earlier authors, had the benefit of hindsight. We all see things now from a different point of view, holding a different set of standards. In a history, we can applaud the far-sighted but we ought not to judge the perceived failings of the time by our standards. Chemical pesticides had enjoyed a generally fine reputation. When doubts were cast upon them, some preventative action was taken quite quickly. Two points that do come into focus with hindsight are that the initial research into the effects of these compounds was not altogether thorough and that much devastation of our wildlife could have been avoided by the swifter introduction of stiffer controls. The comments of the 1960s prob-ably were valid; with hindsight, we see that the failure was compounded in the 1970s and 1980s. The year 1973 saw a supposed final ban on the supply of aldrin and dieldrin. Most uses were halted in 1981, but a total ban on all uses did not come until 1989 and stocks were being used up at least until late 1992.

Why have these controls taken so long to be ratified, after the first flurry of activity in the 1960s? Why was there a need in 1988 to set up the Pesticides Trust to press for tighter controls on these chemicals, in the light of fears, not only for our wildlife, but also for human health? The initial lessons of the 1960s, which were learnt so painfully, were not acted upon. Pressure from other interests was evidently greater than that from conservation. Yet the number of 'unknown' compounds unleashed into the environment was increasing all the time. Only a handful have been mentioned here.

However, if we were wrong in the past to praise the benefits of all pesticides without question, we shall be equally wrong now to damn them all as out of place. Some effects are good; some reactions advantageous to the recipient animals or plants. Like medicine, a little may do the trick; only an excess will cause harm. There is no such thing as a toxic substance; only the concentration is toxic. Many natural elements become toxic at high concentration. Radioactivity is the obvious example. Some individuals of a species can take more than others. That may be a case of the survival of the fittest in the modern world. If so, it may, pervertedly, be creating a stronger population. If, by doing so, it slightly alters the balance of nature, it is merely continuing the process of evolution that has been going on for millenia.

The problem with man-made variations – those that are not sudden or extreme – is that they may have no practical effect on their own, but when other factors, either natural or artificial, come into play, untold damage may be wrought. Polychlorobiphenyls (PCBs) were widely used, not only as pesticides, but also in the production of paints, inks, adhesives and dozens of other industrial processes. They have been found in much of our wildlife since investigations began in 1966. During the autumn of 1969, 17,000 dead seabirds, most of them guillemots, were washed ashore around the Irish Sea. Probably five times as many were never discovered, bringing the toll close to 100,000 birds. They died as a direct result of severe September storms. But many would have survived had they not already been in poor condition due to PCB contamination. Had mankind not had a hand in their condition, some at least would have weathered the storm.

So we must not damn progress and technological advance but neither must we underestimate its potential to disrupt. We must not view it in isolation from all else in the world. We must use our knowledge and ability to give an honest appraisal of anticipated and actual effects. Picking fact from fiction, whether or not it flies in the face of vested interest or conservationist claim, will benefit us all and avoid the sort of emotive statement written in one 'quality' British newspaper in 1967 by a professor of biology:

> Who knows that the crime wave and moral deterioration which is said to be affecting Britain and the world at the present time is not, in fact, the direct result of the widespread use of insecticides, and indeed of other biologically active agents used in connection with food production and storage.
>
> (Quoted by Moriarty 1975)

THE NEW CONSERVATIONISTS

Unresearched and emotive statements claiming that the world's future had been decided and that devastation was about to strike were a sign of the times in the 1960s. 'Doomsday prophets' took the stage. They questioned the basic principles of mid-twentieth-century life and based their questions on that soundest of arguments: that on a planet of fixed size and limited resources, it would prove impossible to realise unending growth. A plethora of plaints and publications hit the headlines with dramatic and chilling assertions on a variety of disasters: mineral failure, over-population, pollution. Many were overstated, some were plain scaremongering, and brought down upon the movement as a whole much otherwise undeserved derision. But most of the derision came from those with vested interests in a solely economic existence. Their ridicule was not well-founded, and the new conservationists – for such they were – parried with much success. They picked holes in the threadbare fabric of economic theory; they halted the time-honoured arguments with new considerations. In the face of derision and an economic inertia which affected us all, the new movement brought the message home so successfully simply because its wisdom was so clear. Of positive action in response, there was precious little, and our general direction has remained largely unaltered. But what the new conservationists did was to open eyes to alternatives, to cause much discomfort of accepted attitudes and to encourage a healthy questioning of the paths we were taking and the goals we sought.

The nature conservation movement has never known such smart success. Perhaps its basic concept is harder to put across. Over-population or failure of mineral resources present themselves as more immediately devastating to human ways than the disappearance of wildlife. Yet the unstinting commitment of the conservationists throughout their longer history ought to have produced a more balanced view. It seems that they have lacked the foresight to relate their beliefs to the everyday situations which involve those with other interests at heart. Often it has been the commitment itself that has blinded the foresight and tunnelled the vision. The arrival on the scene of the new conservationists is a case in point. Nothing, at the end of the day, more fundamentally involves and affects nature conservation than the problems that concerned the new movement. Yet the nature conservationists kept their distance. They were discomforted by the presence of other 'conservationists' whose cause, they seemed to believe, was so far removed from theirs.

To this day, nature conservationists shy away from the thorny issues of energy production and population growth. They actively participate in consumerism and wasteful use of resources. Still they fail to accept the inexorable links between 'the fragments of evidence that told us something was wrong with the world – dead birds, oil in the sea, poisoned crops, the population explosion' (Chisholm 1972).

Almost by definition, the new movement tended to be international in character. Two of its groups in particular became household names owing to their high-profile direct-action campaigns. Greenpeace had its origins in 1969 in a body named 'Don't Make a Wave', but first hit the headlines in 1971. It has been making waves effectively ever since (p. 256). Friends of the Earth (FoE) was founded in America in 1969, the UK branch – which does not, in fact, cover Scotland – taking off two years later, and the Scottish branch in 1979. Throughout its life, it has enjoyed a strong and committed following. Membership of FoE (UK) reached 100,000 in 1989 and 200,000 in 1995, continuing to prosper when other bodies in the environmental movement looked set to stagnate (p. 252).

Plate 6.1 The *Sea Shepherd*, former flagship of Greenpeace supporters engaged
in flamboyant escapades that seized public attention more effectively than
Parliamentary lobbying. Photograph by the author

In their varied campaigns on recycling, dumping, whaling, nuclear
testing and many other concerns, both groups have courted media interest
with their eye-catching, dramatic, sometimes illegal positive action. The
nature conservationists generally chose to remain apart; to fight their own
moderate battles by more acceptable methods; to leave the fundamental
questions to be asked by other groups such as these. They did not wish
to become associated with the law-breaking tearaway image that these
groups engendered. In 1984, the Nature Conservancy Council had to admit
that nature conservation was still seen and practised only as a 'cultural'
activity in Britain. It accepted that only organisations such as FoE and
Greenpeace were tackling the wider 'economic' aspects of the problem.
Lord Cranbrook of CoEnCo voiced the 'establishment' view of these
groups when he complained that they 'exceed the bounds of propriety'
and were motivated primarily 'to enlist support and raise money'. The
latter is something that the nature conservation movement, of which
CoEnCo is part, has always done and will always need to do; the former
is perhaps an effective way of doing it in the mind-numbed modern
world. Were the nature conservationists assisting their cause by keeping
their distance or were they missing valuable opportunities to learn new
strategy?

while the government is disposed to duck difficult decisions ... it is likely to be the potential of organisations like FoE to capture media attention which will keep money flowing to protect SSSIs rather than the NCC's private anguish and entreaties behind the scenes.

(Adams 1986)

Nevertheless, FoE in particular worked hard to throw off its early image of renegade extremism and is now viewed as a sound scientific and political force. Its executive director, Charles Secrett, said on a radio interview on 4 October 1995: 'We are a pressure group that *does* let the facts get in the way of a good story'. FoE's stated aim is to 'transform society' (no less) by:

- changing political policies and business practices to favour environmental protection, conservation and the sustainable use of natural resources;
- empowering individuals and communities to lead sustainable lives, and inspiring them to take peaceful personal and political action to protect threatened nature and conserve natural resources and ecological life-support systems;
- stimulating wide and intelligent public debate about the need and means to encourage environmentally sustainable development.

Another element of the new movement, but one more readily acceptable to existing groups, was the Conservation Society, formed in 1966. Using more traditional tactics, it aimed to make people more aware of the effects of population growth and the unwise use of technology; to stimulate intelligent and humane action to remedy threats to the quality of life, and to promote the conservation of wildlife, natural resources and human cultures. Early campaigns targeted pollution by chemicals and sewage, and unsympathetic building development. In 1970, it formed the separate Conservation Trust to handle the educational aspects of its work, but reformed as a single body in 1987 under the banner of The Conservation Trust. In 1992, it merged with the Environment Council.

NEW SOCIETIES FOR AN OLD MOVEMENT

Rather than widening its outlook to take in the concerns of the new conservationists, the nature conservation movement spent the 1960s filling in the gaps in its network of societies. At a time when the RSPB was entering its eighth decade, there was still no group catering for our sixty species of butterfly, in spite of their familiarity and attractiveness. Some had already been lost from Britain and many others faced extermination under the growing pace of a variety of threats. Most damaging was loss or alteration of habitat which, combined with natural changes in climate and population cycles, was bringing many species to the brink of extinction in Britain. Historically, we had lost a number of butterflies, such as the large copper (p. 29) and the mazarine blue. After the disappearance, in 1925, of the black-veined white – once an orchard pest species – things might have mistakenly looked better because, for fifty years, there were no further extinctions. But modern techniques of land management were taking their toll. The cessation of wholesale coppicing and the introduction of coniferous trees to our woodlands severely affected a number of species, such as the brown hairstreak, purple emperor, wood white and many of the fritillaries. The chequered skipper had disappeared from English woodland by 1976 leaving just a remnant population hanging on in Scotland. The loss of elm through Dutch elm disease devastated the white-letter hairstreak which relies upon it for food. A number of species depend upon the downland habitat; brown argus, small heath, meadow brown, wall and some of the blues are examples. Ploughing of the downs wipes out these butterflies at a stroke.

The drainage of wetlands continues to disadvantage species such as the marsh fritillary, just as it did the large copper 150 years ago. It continues to disadvantage the large copper because, for all the extensive management effort at Woodwalton Fen, a continuing drying out of the area due to the arable farming that surrounds it still prevents the establishment of a viable population. In the uplands, afforestation is driving out the large heath. The grayling and the silver-studded blue are disappearing with the heathlands and, in 1979, modern techniques of grassland management caused the loss of the large blue – the first extinction for over half a century. One-third of butterfly species are currently in serious decline, with the high brown fritillary holding the dubious position of being the most endangered.

Compared with habitat destruction, other threats are not regarded as serious. Pesticides have effect on butterflies in the sprayed areas but not much further afield. Collectors, as always, may represent the final straw of extinction but do not cause healthy populations to decline. Natural fluctuations in numbers and climatic variations are almost impossible to cater for.

It was as a result of increasing concern for our butterflies that, somewhat belatedly in 1968, two amateur naturalists, Thomas Frankland and Julian Gibbs, formed the British Butterfly Conservation Society (BBCS), since shortened to Butterfly Conservation. Its aims are to:

- save from extinction or protect all species of British butterfly, by conserving them in the wild by such means as are available or by breeding numbers in captivity and, where practicable, reintroducing them in natural habitats;
- sponsor further scientific study and research in conservation of these butterflies both in the wild and in captivity;
- foster interest generally by educating the public, and in particular

educational establishments, in problems concerning conservation of these butterflies.

The young society worked hard to highlight the threats but its early growth was slow and membership only cleared 9,000 during 1992. It assumed responsibility for the Butterfly Monitoring Scheme from Monks Wood Biological Records Centre in 1983, and three years later, at the invitation of the NCC, became actively involved in the work with the large copper at Woodwalton Fen. Its first reserve was acquired jointly with the Worcestershire Trust for Nature Conservation in 1986. Of more importance has been its imaginative work on habitat management in the wider countryside which does, after all, present the best hope for the future of our butterflies. For the threats remain immediate and real and public concern remarkably dilatory in comparison, say, to the enthusiasm shown for conserving birds.

The Butterfly Protection Committee of the Royal Entomological Society, which later became its Conservation (Insect Protection) Committee (p. 55), endured another heavy change of name in 1968. This time it became the Joint Committee for the Conservation of British Insects (JCCBI). The RES felt that the base of insect conservation needed to be widened and co-opted onto the Joint Committee members of the BBCS, the British Entomological and Natural History Society and the Amateur Entomologists' Society. Observers were invited from the NCC, the FC, the Agricultural Development and Advisory Service, the Ministry of Defence and the National Trust. The Committee monitors insect conservation in Britain, supports the international movement for insect conservation and acts as a forum for discussion on rare and threatened insects. The JCCBI is purely voluntary, relying on funds from its constituent bodies for day-to-day running and from the likes of the WWF for specific survey work. It is acutely aware of its small size in the conservation movement and actively seeks to increase its ability to extend the vital work that it performs.

Another newcomer to the network of societies was Philip Glasier's Hawk Trust (now the Hawk and Owl Trust) of 1969. In the face of grave concern over the effects of pesticides, it was dedicated to the appreciation and conservation of all our birds of prey. Glasier, however, was best known as a falconer, and that angle of interest amongst the membership attracted the suspicion of other conservationists. Although it has been a popular practice for many centuries, the captive keeping of raptors still upsets the sensibilities and awakens the wrath of some conservationists, and questions are often asked regarding the source of the falconer's birds. However, over the years, the Hawk Trust has worked hard to show its goodwill in traditional conservation. As a society, it no longer breeds species in captivity, but concentrates on wardening them in the wild and carrying out survey work. With growing acceptance by conservationists, it now sees itself primarily as a mediator between them and the landowners and sportsmen and it does much to encourage a conservation-minded approach to land management, particularly on the sporting estates.

Also concerned with sporting matters is the Game Conservancy. This independent body developed in 1969 from a game research institution, run by ICI since the 1930s, and the Game Advisory Service of Eley Cartridges! Thus it started with a membership of 3,000 which had almost tripled by 1980 when it became a registered charity, The Game Conservancy Trust. At this point, the advisory functions on sporting matters devolved to a separate Game Conservancy Ltd. The Trust concentrates on research and education in the conservation of game species such as partridge, pheasant, hare and roe deer, and carries out an annual national game census.

The development of both the Hawk Trust and the Game Conservancy Trust indicated the continuing distrust between sportsman and nature conservationist. Why else the need to shelve the sporting angle of the Game Conservancy to a separate body? Why the need for the Hawk Trust to act as a buffer between the two sides? How much healthier it would have been to see those similar countryside interests merging to fight their battles in unison. But the bridges had still to be built.

THE DEER AND COUNTRYSIDE ACTS

One of the things held in common by sportsmen and conservationists was the wish to maintain populations of deer at a high and healthy level. The primary association concerned with deer is the British Deer Society. By encouraging both sportsman and conservationist to its membership, it is one of the few bodies that provides a credible forum for the two sides.

Red and roe deer are natural to Britain; the fallow was introduced by the Normans; the sika, followed by the muntjac and Chinese water deer, are late-nineteenth-century introductions. During the 1940s and 1950s, concern had been growing over the size of the deer stock in Britain. In 1959, the Deer (Scotland) Act brought into effect a number of measures against poaching and a close season for red deer: 21 October to 30 June for stags and 16 February to 20 October for hinds. It also established the Red Deer Commission, an independent body appointed by the Secretary of State for Scotland. The commission is concerned with all aspects of the conservation and control of red and sika deer and to a lesser extent, of roe deer. In 1966, protection was extended to all deer in Scotland. The Deer Act 1963 makes provision for red, fallow, roe and sika deer in England and Wales. Here the close season is 1 May to 31 July for stags and 1 March to 31 October for hinds. Dates may be varied by the Secretary of State. It is an offence to take deer at night or to set any sort of trap for them.

The legislation proved effective, for deer now pose a significant threat to woodland management and restoration. In some regions, they suffer the effects of over-population. Between the mid-1960s and the mid-1990s, numbers of red deer in Scotland doubled to reach 300,000, and active steps were being taken to increase culling in order to cut the population by one-third. There were calls for the withdrawal of the close season on sika and red/sika hybrids, and for the Red Deer Commission to be able to authorise shooting out of season in order to protect the natural heritage. Habitat destruction has not proved a problem for deer. They survive in a variety of habitats, including towns in the case of the smaller species. Afforestation has encouraged some return to their original habitat. New forestry attracted roe deer to return from Scotland to some former haunts south of the border.

Efforts to force protection in law for other types of wildlife, particularly the plants, failed during the 1960s (p. 113), and the other notable legislation of the decade, the Countryside Acts, related mainly to access. A growing awareness of the natural environment, coupled with the easy optimism of the 'swinging sixties', was bringing far greater numbers of people out into the countryside. A new spectacle evolved in the landscape; the nature trail and the picnic site. In 1963, one of the first nature trails was opened at Coombe Hill in Buckinghamshire. The idea fired the public imagination and, during the second National Nature Week in 1966 (p. 114), the NC and many voluntary bodies cut the tape on a number of new trails.

Since the Forestry Act of 1951, the FC had been able to respect the amenity potential of its land when considering policy and planting programmes. Sylvia Crowe, the commission's own landscape consultant, published *Forestry In The Landscape* in 1966. It called for improvements in the landscaping of plantations and for increased leisure use of our forests. As a result, new areas were made available for public enjoyment. In 1964, there were no formal picnic sites on FC land. By 1969, there were 104, together with ninety-two forest trails, and Darling (1971) was able to say that, 'since those pioneer days, there has been a drastic change of outlook. The commission is in the forefront of dedication to the philosophy of multiple use'.

On the commons, the picture was not so rosy. The Royal Commission on Common Land had been set up in 1955 as a result of difficulties in returning to the commoners land which had been requisitioned during the war. Its report of 1958 called for full public access to all commons and an emphasis upon recreation. In its view, management schemes ought to revolve around the needs of the commoners and the public. Many landowners fought against this concept for, although much common land was used by the public, legal access was restricted to only 20 per cent of English commons. Farming and forestry interests were particularly concerned, as undoubtedly were many nature conservationists. Those commons for which owners could not be found were to become a public responsibility. Thus the commission called for the registration of all common land, an idea first mooted in the 1942 Scott report. This was to be followed by a Common Land Act which would outline the rights of owners, commoners and the public. The Commons Registration Act 1965 ruled that any land not registered as Common Land or Town or Village Green by 1971 would be legally 'deregistered'. Unfortunately, the Act did not specify who was to do the registering and so it was that, right through to the present, disputed claims are still being taken to the commission. Meanwhile, the High Court has been bound to find that any land not registered is no longer common land and much is thus threatened by ploughing or planning proposals (p. 177).

By the mid-1960s, it was clear that some official machinery for country-side recreation was badly needed. Even the National Parks Commission was faltering. In 1964, its duties were transferred to the Ministry of Land and Natural Resources and, in the following year, plans were announced for a new commission with powers over the wider countryside.

Northern Ireland was first to get the matter onto the statute books with its Amenity Lands Act of 1965 promoting recreation and conservation. The 1967 Countryside (Scotland) Act and the 1968 Countryside Act (covering England and Wales) introduced the promised Countryside Commissions. These purely advisory bodies sought to encourage public access and enjoyment of the countryside, but were also to have regard for the conservation and enhancement of natural beauty. The National Parks Commission became the Countryside Commission (CC) on 3 August 1969 with duties towards recreation in all areas of England and Wales. Inevitably, the National Parks continued to suffer under this wider remit so, in 1974, the National Park authorities took over their administration and, three years later, the Council for National Parks (p. 78) was given an independent remit to guard their interests.

In Scotland, of course, there had been no National Parks Commission, no official countryside watchdog to guard against damaging development. It is hard to believe that, until as late as 1967, only a handful of voluntary bodies with relatively small memberships was doing anything to champion the conservation of our most spectacular countryside in the face of planning development. The Countryside Commission for Scotland (CCS), established by the 1967 Act, took on the prodigious task of encouraging conservation and recreation over 98 per cent of the country.

In spite of its name, the 1949 Act had achieved less for recreation than it had for nature conservation. Conversely, the Countryside Acts intro-duced more measures to please the amenity lobby than it did the conser-vationists. For they went much further on behalf of recreation than the mere establishment of the Countryside Commissions. They broadened the scope of access arrangements – which had only applied to mountains and moorland under the 1949 Act – to take in woodland, rivers and canals. But their most notable contribution to recreation was an entirely new idea and one which has come to play a far greater role than was expected. The country park was envisaged as 'an area of land or land and water, normally not less than ten hectares or more than 400 hectares in extent, designed to offer to the public, with or without charge, opportunity for varied recreational activities in the countryside'. The Countryside (Scotland) Act 1967 defined it as 'a park or pleasure ground in the countryside which by reason of its position in relation to major concentrations of population affords convenient opportunities to the public for the enjoyment of the countryside or open-air recreation'. John Cripps, Chairman of the CC, was at pains to point out that they were not intended to be amusement parks. Their aim was to provide 'more informal countryside recreation, primarily on a family basis . . . some quiet enjoyment of the countryside and nothing more'.

Country parks have little in common with National Parks. They are designated by the local authority and overseen by the Countryside Commissions. They are small pockets of tame, organised countryside, close to centres of population. They cater primarily for the less energetic and the day-tripper; for the 'short run into the country'. Consequently, many centre on large houses, reservoirs or other man-made features. With certain valuable exceptions, they generally attract little wildlife, and what there is must often be called to attention by nature trails and brochures. In this respect, the majority of country parks have done little to further the cause of nature conservation. What they have more importantly done is to relieve much visitor pressure from the wilder, less accessible areas, by presenting opportunities for open-air relaxation closer to home. For instance, the country parks at Llys-y-Fran and Shotton Manor in Dyfed intercept the people of industrial South Wales before they reach the Pembrokeshire Coast National Park.

The confirmed ramblers are suspicious of this sort of enterprise. 'The Countryside Commission, with its policies of shepherding people away from the wild parts of Britain, is undermining the spirit of adventure, which has been an essential part of the human race since the beginning of time' (Hill 1980). But it is, after all, only a policy of shepherding. Nobody who wishes to seek out the wilder countryside is forced into the country parks and, when they reach their goal, they will find greater peace and solitude for the existence of those smaller parks. Hill's complaint is based upon a broader dissatisfaction: 'the refusal to face up to the controversial issue of the desire to roam freely over all uncultivated land'. It is indeed a controversial issue and a source of countryside conflict on a par with field sports. Whichever way it is looked at, free access to all wild country would be detrimental to our wildlife. The rambler, even the naturalist, would gain – for a short time – but the conservationist would not. 'The most worrying proposal to appear in the Peak District National Park Plan for 1978', as far as Hill was concerned, was the suggestion that access to the moors be limited to defined paths, during the breeding season only, for the benefit of nesting birds. 'Sheffield ramblers are refusing to accept these new restrictions . . ..' Until selfish and short-sighted statements such as these make way for open discussion towards an integrated agreement, the users of the British countryside will continue in their failure to achieve their separate aims in the face of stronger, more united opposition.

We must not derogate initiatives such as the country parks. The fact that thousands of people are content to enjoy a tamer type of countryside is a straw to be clutched and worked to best effect. By their level of success, the country parks proved that they were needed. The earliest ones were at Elveston Castle in Derbyshire, Wirral Way in Cheshire and Culzean in Ayrshire. From these, the concept grew beyond all anticipation and, by 1995, there were nearly three hundred country parks in Britain.

Of rather less effect were the provisions of the Countryside Acts on behalf of nature conservation. Steps were introduced to prevent damage to moorland in the National Parks. The minister could make an order requiring a landowner to give six months notice to the local authority of any intentions to plough heath or moorland which had been out of agricultural use for at least twenty years. The NC was given leeway to make management agreements with the owners and occupiers of valuable sites not designated as statutory reserves. It was empowered, for the first time, to compensate landowners who would agree to continue farming by traditional methods for any potential loss of yield. During the 1980s, this became a normal practice but in 1968 the notion struck fear into the hearts of conservationists who had somehow never envisaged having to pay hard cash for their cause. In the event, the landowners showed no interest in the scheme at that time.

Rather more popular with the conservationists was Section 11 of the 1968 Act which required all government departments to 'have regard to the desirability of conserving the natural beauty and amenity of the countryside'. Conversely, Section 37 placed a duty upon all ministers, the CC and the NC to have regard to agriculture, forestry, economic and social interests. Such clauses were nothing more than political gamesmanship. The ministers showed as little interest in conservation as the farmers did in compensation, and nothing changed. Having regard to interests outside the scope of one's own department was hardly likely to affect the activities of a department. Not until the end of the 1980s, with the inception of Environmentally Sensitive Areas and the release of some land from farming, could MAFF and the NC be imagined to be working together.

DEFEATS AND SUCCESSES

Yet the very inclusion of Section 11 shows that the conservationists were making themselves heard and that the statutory lip-service of the 1950s was at least still being paid. The nature conservation movement had suffered some crushing defeats in recent years. That was nothing new, but the publicity that surrounded them was, and the politicians were having to take note as a result. The Countryside Acts came within months of the royal assent granted in March 1967 to the giant ICI company to flood over 300 ha of the upper Teesdale valley in the Pennines. The reservoir was required to supply water to the chemical works at Billingham on Teesside. The site chosen was rich in rare and local arctic and alpine flora, unchanged since before the post-glacial forest cover had wiped it out in other regions: plants such as mountain avens, rock rose, thrift and sea plantain. Nowhere else could such a special plant community be found. 'The fight to stop the intrusion was dramatic and involved virtually all the important figures in conservation' (Dawson 1980). A Parliamentary Select Committee was set up to hear the arguments. For all that, Cow Green reservoir was given the go-ahead. But the industrialists had been taken aback by the ferocity of conservationist feeling and so too had the politicians. Within three years, the two largest political parties had appointed their first Ministers of the Environment.

Mabey (1980) quotes from Lord Kennet's comments in the House of Lords; he saw the granting of permission for Cow Green:

> as the moment when the British Parliament accepted to the full its duty to examine ... the fundamental conflict of interests between one ponderable – industrial and economic progress – and two imponderables – pure research and the preservation of natural beauty.

Cow Green had been all about plants. Less dramatic but more sinister was the continuing failure of any measure of plant protection to reach the law books. It must be a fundamental concern of conservationists why any particular type of threatened wildlife should not receive protection when a campaign is set in motion on its behalf. Why was it that the British had legislated against the taking of a single sparrow's egg and not against the complete extermination of a rare plant? There was grave concern in the 1960s over the excessive picking of some plants and the ineffectiveness of local by-laws to control it; over the commercial taking of moss, roadside spraying, and the problem of keeping secret the locations of rare species. Consequently, in 1963, the BSBI, together with the SPNR and the Council for Nature set up a Wild Plant Protection Working Party which, four years later, introduced a Conservation of Wild Flowers Bill. This sought to outlaw the picking, uprooting, and offering for sale of any species on a schedule, and the uprooting of any plant without the landowner's consent. It failed, and a simpler Bill was quickly drafted to illegalise the picking or uprooting of some twenty rare plants and the offering for sale of a further forty species when picked without the landowner's permission. Sadly, this compromise failed too and the plant protectionists had to wait until 1975 before they got anything onto the statute books (p. 148). The 1960s closed with plant protection unfulfilled except for some ignored and outdated county by-laws and the voluntary movement doing what it could to encourage consideration for our plants by issuing leaflets and codes of conduct.

Yet the nature conservation movement did score some successes and perhaps felt that its message was beginning to infiltrate the national conscience. There were a handful of highly successful campaigns. The week of 18–25 May 1963 was National Nature Week, sponsored by the Council for Nature. A National Wildlife Exhibition was held in London and commemorated with a special set of postage stamps. Other displays and activities throughout the country, largely run by the local trusts, awoke much public interest and brought some useful contact between landowners and conservationists. The exercise stimulated not only a second National Nature Week in April 1966 but also, and more importantly, the highly successful 'Countryside In 1970' conferences (p. 121).

Another campaign to catch public attention was the National Trust's Enterprise Neptune, launched in 1965. A coastal survey completed by the trust two years earlier had found that, of 4,800 km in England, Wales and Northern Ireland, one third had been spoilt beyond redemption, another third was of no particular interest and that, at best, about 1,450 km were worthy of urgent protection. Concern was widespread – as was the evidence to justify it – that the planners were failing to cope with the detrimental effects of increasing leisure time on our coastline. The campaign's aim was to take control of almost 1,600 km of coast, including the 300 km already protected by the trust. Public response was good, with over £500,000 raised in a year. The first acquisition was at Whiteford Burrows on the Gower Peninsula. In November 1973, Neptune's original target of £2 million was realised. But the NT now estimated that over 3,000 km of coastline were beyond redemption and boosted the campaign for 'The Next 100 Miles'. A separate campaign was mounted by the National Trust for Scotland.

The most impressive piece of fund-raising machinery in the nature conservation movement dates from 1961. In its first twenty-five years, the World Wildlife Fund (WWF) raised over US $110 million. More than half this figure came in during the final five years, demonstrating the fund's continuing strength. In that quarter-century, organisations in twenty-three countries spent almost £40 million on more than 4,000 conservation projects in 130 different countries. 'World Wildlife Fund: An International Foundation for Saving the World's Wildlife and Wild Places' was set up by the poorly funded IUCN at the instigation of a group of eminent British conservationists. The British national appeal began on 23 November 1961. The WWF giant-panda logo is now familiar throughout the world. Its Charter is less familiar:

> It is the responsibility of all who are alive today to accept the trusteeship of wildlife and to hand on to posterity, as a source of wonder and interest, knowledge and enjoyment, the entire wealth of diverse animals and plants.

This generation has no right, by selfishness, wanton or uninterested destruction, or neglect, to rob future generations of this great heritage. Extermination of other creatures is a disgrace to mankind.

The fund, now known as the World Wide Fund for Nature to reflect its concern with the total environment, is the world's largest conservation body with active national appeals working all over the globe.

Campaigns such as these fired public imagination and helped the conservation message to seep into the national conscience. On 9 October 1961, in preparation for the launching of the WWF in Britain, the *Daily Mirror* newspaper ran a 'shock issue' which highlighted in seven full pages some of the dangers facing our wildlife. The sum of £35,000 was donated as a result. Media interest in conservation, minimal before the mid-1960s, increased threefold between 1965 and 1973, as stories worthy of publication began to multiply. Take, for instance, the rejuvenation of the country's premier river. In 1961, flocks of over fifty ducks were sighted on the Thames between London and Swanscombe for the first time since the turn of the century. Teal, pochard, shelduck, and the first fish were recorded at Tower Bridge in 1963. The year 1968 saw terns at West Thurrock, smelt at Fulham and bass at Blackwall Point. By 1971, fifty-four fish species could be found in previously empty stretches of the Thames, and West Thurrock hit the news again in 1974 with the first salmon for 100 years. Discharges of warm water from the power stations, and of chemicals, were being reduced and new sewage treatment works were in operation at Beckton and Cross Ness. The nation was witnessing the recovery of its capital river from debilitating pollution – and that was news indeed. Other rivers followed suit, but more reluctantly. I remember seeing soap suds scudding across the River Trent at Nottingham in 1972. Another ten years were to pass before that river reached today's questionable standard.

The news item of the decade came in 1967. This was the year that witnessed the introduction of a ban on the dumping of waste oil at sea. But bans and conventions are powerless against accident. On 18 March 1967, the 120,000 tonne tanker *Torrey Canyon*, overladen by 1,000 tonnes of crude oil and sailing under the Liberian flag of convenience, was 10 km off course when it ran aground on the Seven Stones, Isles of Scilly. Over half its cargo was immediately released into the sea, forming a slick which, within twenty-four hours, measured 20 km by 10 km. All parties, to their continuing disgrace, were very slow to act and much prevarication over likely costs went on while the ship broke up. Ten days later, the Royal Navy was called in to bomb the wreck and burn off the remaining oil. The toll on wildlife has been well documented but never fully quantified. All marine life was drastically affected but the pictures that reached the public most often were of the seabirds. They died of exposure as the oil ruined the waterproof nature of their feathers, or of the effects of ingesting the oil as they tried to remove it.

Pictures of these crippled creatures and of blackened beaches brought home overnight the consequences of oil pollution at sea . . . and the biggest iniquity of all was that we did not know how to deal with it. The trade in oil had been going on across the world for decades, and now we discovered that there were no contingency plans, official or otherwise, to cater for what was, after all, eventually inevitable. Thus the ten days of prevarication while the problem grew worse. Thus the use of inappropriate and damaging detergents to clean the beaches and the birds. Some £1.3 million was spent on detergents; more than the total annual grant to the NC at that time. But those detergents caused cell damage and death to millions of marine creatures and did little good for the birds. Research has since shown that more suitable compounds were available.

The conservationist is wrong to damn technological progress but right to insist that we should first be clear about our actions in the event of disaster or disruption. Without these plans, we abrogate our responsibilities.

Fiascos such as the *Torrey Canyon* episode were beginning to show the warnings of the conservationists in a different light. One or two shifts towards their way of thinking could be discerned. In 1969, thanks to the efforts of the county naturalists' trusts, twenty local authorities agreed to time their cutting of verges to minimise disruption to wildlife, and many of them agreed to limit their use of herbicides. The monetary savings quickly became clear and the authorities needed little further encouragement to allow a natural improvement of this habitat. With a total of well over 200,000 ha of verges in Britain, an area the size of Cambridgeshire, this represents a valuable wildlife resource. Twenty of our fifty or so mammals, forty of our two hundred birds and all six of our reptiles have been found breeding in roadside habitats, as have twenty-five of our sixty butterflies and eight of our seventeen bumble-bees! Four hundred species of plant were discovered on a random survey, suggesting that perhaps a total of 650 flourish on our roadsides. These figures are from Ratcliffe (1977) who suggests that, on motorway verges in particular, 'there is a very real chance to practise creative conservation'. With our road network still expanding, this is a chance not to be missed. Conservationists were quick to act in the early 1970s when a colony of rare columbine was threatened by the construction of the M1 motorway. Members of the Nottinghamshire Trust for Nature Conservation collected seed from the plant and propagated over eighty seedlings which were later transferred to the new motorway verges.

The FC, as well as being more amenity-minded, was also becoming committed to general conservation ideals. By 1962 it had increased the wooded area of Britain from 1.1 million ha in 1914 to 1.7 million ha and had control over further unplanted land. Its *State Forest Memorandum* of June 1965 was therefore encouraging to conservationists for it based its policy towards animals and birds upon three assumptions:

- that, while it must prevent damage to forest crops and neighbours' property by animals living in the forest, there should nevertheless be no indiscriminate wholesale destruction of any species;
- that it had responsibility to conserve and manage wildlife by virtue of the forests' acting as a wildlife reservoir;
- that there should be no collecting of any species or any eggs.

There were moves afoot even in political circles. During the pre-election months of 1970, the Labour government published a White Paper on *The Protection of the Environment: The Fight Against Pollution*. In its discussion of pollution ranging from chimneys to sonic booms, from air to river pollution, from control to penalties, the paper was the first ever to be written from an ecological standpoint.

NOT ENOUGH PEOPLE

In all sorts of ways, in all walks of life, the message of conservation was beginning to niggle the national conscience. Yet the movement could hardly claim any real success. Enterprise Neptune had to be placed alongside the *Torrey Canyon*; sympathetic sections of statute alongside Cow Green; a reduction in roadside spraying alongside the failure to protect our plants. The nature conservationists were moving towards that seemingly unshakable situation. For all their growth as a movement, they were not ringing the changes in the nation's outlook. Basic aims and tactics needed to be reviewed. They had known it for years:

> bird protection can never be really efficient until sanctuaries can become national parks, individual problems become local problems, and local problems become national problems. We cannot really know what is to become of our birds, and from this knowledge manage them, until we treat them like any other national commodity such as sewers or the unemployed or electricity.
>
> (Fisher 1940)

> Nature conservation, if it is to be successful, must form part of a national policy which includes agriculture and industry, urban development and transport.
>
> (Vesey-Fitzgerald 1969)

Thirty years apart, but the message remained the same. Little had been done to attract the sort of high priority that was vital for success. Sights had continually been set at a low level: abide by the Country Code, report offenders against the law, channel the interests of youngsters, join your local naturalists' trust, encourage wildlife to your garden. With this sort of modesty, the movement would never make the grade.

Formal education was doing nothing to encourage interest in natural history, let alone its conservation. Some scientific ecologists still remained unconvinced that mankind had any relevance to their minute studies of wild communities. Nicholson (1970) quotes from a Ministry of Education document of 1960:

> The place which is occupied by advanced biological studies in schools ... is unfortunately that of vocational training rather than of an instrument of education ... the contribution which the study makes to the pupil's education is so small that it is doubtful whether such a subject ought to find a place in a school at all.

This quite astonishing admission of official failure must nevertheless be seen as a reflection of the conservationists' lack of vigour in promoting their message where it would be most effective. Yet there were societies for the younger generation: the RSPB's Young Ornithologists' Club, formally known as the Junior Bird Recorders' Club; the XYZ Club of the Zoological Society of London; and the British Junior Naturalists' Association. But such societies were preaching to the converted; most of their members joined of their own volition as a result of an interest they already held. The same can be said, of course, of the movement's adult membership.

Another problem, foreseen but postponed, was that, although nature reserves had much to commend them, they could not survive in isolation. They and their wildlife depended upon the conservation of the wider countryside in which they were stranded. Lord Hurcomb, in his president's speech at the RSPB's seventy-fifth anniversary dinner in 1964 said:

> We have come to realise that to protect natural life it is not enough to protect a particular species by preventing it from being shot. We realise that if wildlife of any sort is to persist it must be given a proper habitat.

In the following year, the NC initiated a national survey of sites whose preservation was regarded as a matter of urgent priority. They were selected with a view to preserving all types of habitat, if not in designated reserves, then as part of the working landscape, and were eventually published in Ratcliffe (1977). The BTO began a similar survey of habitats important to birds in 1973.

> The world has awakened to the main ecological fact, that protection of any one creature or a complex of species depends primarily on persistence and survival of the habitat. That is conservation and it applies the world around.
>
> (Darling 1971)

For the conservationists, embroiled in their own interests, the nature reserve was a comfortable arrangement. The whole reason for its existence was to protect wildlife. Nobody else need be bothered. But now the movement was hoping to protect the wider countryside. This would involve the whole population, those very people whom the conservationists had never been able to reach. The task was a prodigious one:

The future of the wild life of Britain does not lie in a few reserves, but in man's attitude of mind.

(Vesey-Fitzgerald 1969)

Without a change of tactics, the nature conservationists could not hope to maintain credibility any longer, not even within their own ranks. The defeats were becoming more drastic and more regular and, ironically, the public outcries which they provoked seemed to reflect badly upon a movement which was apparently failing to do its job. Many who heard the nature conservation message were sceptical about the ability of those who promoted it. They seemed always to stand against specific proposals, as if taken by surprise that such things should be considered. They appeared not to have any constructive policy which allowed for such developments. If they had, it was not being presented. Consequently, they generally lost the case and came across as being negative in outlook. The new conservationists, using more extravagant tactics to tackle wider environmental issues, were dissolving public indifference and official inertia in a much more positive fashion. Certainly, it is they, and not the nature conservationists, who are remembered in hindsight from the 1960s.

Plate 6.2   The future of wildlife lay, not in reserves, but in an attitude of mind, yet the conservationists were maintaining the barriers and keeping the secret to themselves. Photograph by the author

The nature conservationists had now to act upon their own message. Other 'national commodities' were not about to come to them. Firstly, they needed to approach other countryside interests:

> a new large effort was demanded in Britain to break down the barriers between different interests in what came to be conveniently lumped together as 'the countryside', and to bring together a far more broadly based, tolerant and constructive alliance.
>
> (Nicholson 1970)

They would then go to the people of Britain, sure in the knowledge that their message, realistically styled, would be kindly received.

> The urge is there all right, and a lot of first-class scientific data, but not enough people, either in or out of power.
>
> (Darling 1971)

If only the message could be passed to the people, those responsible for national policy would certainly take it into account. But people in general were blinded by the euphoria of the 1960s; unwilling to admit the need for a change in outlook. The conservation movement had to supply the jolt from inertia; it must make nature conservation meaningful to the man in the street. The 1970s became a series of campaigns and special 'years' aimed at bringing the message home to the people of Britain.

> The struggle broadened out, from the restraint of particular abuses and environmental injuries, into a campaign for the hearts and minds of people world-wide, and for profound changes in perceptions and lifestyles.
>
> (Johnson 1983)

7   1970s: Going public and getting places

GOING PUBLIC

European Conservation Year 1970 (ECY) was the closest the nature conservation movement had yet come to embracing all aspects of conservation, and it opened the campaign for the hearts and minds of the people. It was the culmination of the 'Countryside in 1970' conferences, sponsored by the Council for Nature, the Nature Conservancy and the Royal Society of Arts.

The coming together of various factions of the movement during National Nature Week in May 1963, and the level of public interest that arose from it, intimated to the nature conservationists the potential to be gained from closer ties within their ranks. As a result, the Council for Nature convened the first 'Countryside In 1970' conference in November 1963, with HRH Prince Philip as President. Its primary task was to consider the pressures upon land in the UK and to answer the question: 'What sort of countryside do we all want to see in 1970?' Over 200 representatives from ninety interested bodies aired their differences and came to agree that they should encourage a more conservationist attitude in the British public.

The conference set the scene for regular contact, not only between the conservationists and others, such as industrialists and architects, but also and more tellingly between the various conservationists themselves. A dozen study groups were formed to look at a wide range of land-use issues, which then composed the agenda for the second conference in November 1965. Planning and urbanisation, outdoor recreation, land reclamation, law, new technology, the preservation of national riches, both natural and man-made: these were some of the topics that came under scrutiny. The conference recommended the establishment of a Countryside Commission and the sanctioning of local authorities to acquire land for amenity and recreation. These proposals came to be embodied in the Countryside Acts (p. 110).

But a suggestion that county councils set up countryside committees to create nature reserves and encourage conservation projects did not meet with success. Under the Open Spaces Act of 1906, parish councils already had the power to acquire waste or unoccupied land, as well as land less than 20 per cent developed. But there were no guidelines by which they might designate such areas as nature reserve. Over ten years later, the battle was still being fought. At the third reading of the Local Government (Miscellaneous Provisions) Bill in 1976, the Earl of Cranbrook moved an amendment to give parish and community councils power to provide nature reserves. It was defeated by three votes.

By 1970, certainly, 'nothing significant had changed for habitat protection' (Adams 1986). So it was that the third and final conference in October 1970 sought ways in which to involve the wider public in the cause of conservation. A total of 335 organisations were represented, almost four times the number at the opening conference. Again, HRH Prince Philip acted as President. As well as considering agriculture and forestry, urbanisation and leisure, they looked at the prospects for individuals and associations wishing to further the cause; at responsibilities for the environment; at social issues and at 'choice and opportunities'.

The main achievement of these conferences was to bring together complete strangers all of whom had the good of the countryside at heart. The greatest success came not in concrete reform but in the opening of eyes and ears to the views of other interests. In order to consolidate this new-found awareness, the Committee (later Council) for Environmental Conservation (CoEnCo) was formed in 1969 to foster and promote a common approach to major environmental issues. Late in 1988, it changed its title to The Environment Council. It is an independent national charity which co-ordinates the work of the voluntary bodies, acts as a resource centre for them and provides a forum for discussion. The wide span of its bridge-building activities can be seen from its list of members which includes the National Society for Clean Air, the Confederation of British Industry, the Pedestrians' Association for Road Safety, the Civic Trust and the Council for British Archaeology, as well as the main conservation groups. The environmental reports produced by CoEnCo during the 1970s covered such matters as transport, urban pressure, energy and industry. Its ability to come up with balanced comment on topics which overspill the capacity of any one particular interest has earned it much respect in conservationist and political circles.

Convinced that the potential existed for a following which might number millions, the conservation movement 'went public' in 1970. European Conservation Year took the message to the people in a popular and down-to-earth fashion. It was a year of exhibitions, open-days and meetings; of lectures and films; of field trips, nature trails and practical projects; of publicity and media coverage. The UK Provisional Booklet, distributed prior to the event, stated the message quite clearly:

EUROPEAN CONSERVATION YEAR

In 1970 twenty-one countries from Iceland to Turkey will join together in a great co-operative effort to get people to care more for their environment. The outcome is European Conservation Year organised under the guidance of the Council of Europe.

Why is it so important and why should European Conservation Year deserve wide support? The trees and plants which cover the land, and the rich variety of wild creatures living on it, are a basic resource. Without them our health and happiness and that of our children would be diminished. We must use this natural heritage wisely.

The countryside has to serve many needs: as a place to live in, to work in and to play in. Industry wants part of it for factories; farmers use much of it for growing food; foresters need it for timber; and towns-people want it for recreation. All these demands must be fitted in without destroying or spoiling our environment. This is what conservation means.

Aims

The aims of European Conservation Year are:

To seek agreement on policies to conserve and improve the environment and devise the best ways to implement them.

To spread information and education so that citizens in the countries taking part will understand the problems better and support measures to deal with them.

The Problems

European Conservation Year will open with a conference in Strasbourg in February 1970. Delegates from the countries taking part will discuss four major impacts affecting the environment. These are: Urbanisation; Industry; Agriculture and Forestry; and Leisure pursuits. The conference will aim to produce a Declaration which will go to each government and to local authorities in each of the countries to act upon.

Nature conservation had come of age. Overnight, it became an acceptable pastime to watch the birds; normal behaviour to show concern over the plight of a rare reptile; frowned upon to uproot wild flowers. The media were full of conservation stories; books and magazines reflected the new flavour of the times. Golden eagles, right on cue, returned to breed in England for the first time in over a quarter of a century. Ordinary people came up with extraordinary ideas, like the lady who raised over £1,200 for her local naturalists' trust by 'selling off', at a penny a leaf, pictures of trees printed on card. New initiatives abounded. HRH Prince Philip presented Countryside Awards to 100 projects instigated in England in the previous five years and 'worthy of emulation'. Twenty-eight Prince of Wales Countryside Awards went to deserving projects developed in the province during 1970.

It is difficult to imagine that, before 1970, conservation was a totally alien idea to the majority of people. ECY instilled in thousands a concern for at least some aspects of the cause. Many others who remained unmoved were nevertheless made aware of the basic principles. ECY remains one of the most successful of the designated 'years' and it set the pattern for the rest of the decade which proved, in many respects, to be the most successful for the movement.

NEW GROWTH

Perhaps the most obvious aspect of this success was the vast growth in support for the voluntary bodies. The RSPB had recruited its twenty-five thousandth member on the final day of 1964. By 1968, membership had reached 41,000. There was then a sevenfold increase in this grand total during the 1970s, giving over 300,000 members by the end of the decade. With this sort of support, all aspects of the Society's work flourished. Reserve holdings of 6,500 ha at thirty-six sites in 1969 doubled to 13,000 ha at forty-six sites by 1975. Then came the launch of a £1 million appeal, 'Save A Place For Birds'. By 1979, land holdings had rocketed again to 32,000 ha.

The SPNR played a major part in ECY and subsequently reaped the benefits. It launched its *Conservation Review* in the autumn of 1970. It was soon felt, however, that a new image was needed to keep the Society abreast of the times. A revised royal charter introduced the Society for the Promotion of Nature Conservation (SPNC), officially inaugurated on 10 July 1976. *Conservation Review* was given a facelift, and a new badger logo was devised. The aims of the Society reflected the broader view that conservation now went beyond the bounds of nature reserves. Primarily by assisting the county trusts, SPNC was to:

- establish nature reserves to safeguard scarce habitats or to provide 'reservoirs' for rare species;
- disseminate information to local authorities, landowners and others concerned in countryside management;

and, true to the aims of ECY:

- promote the conservation of nature for the purposes of study and research and to educate the public in the understanding and appreciation of nature, the awareness of its value and the need for its conservation.

Each of forty county trusts was granted corporate membership and was represented on council. They included the Scottish Wildlife Trust, formed in 1964 in the absence of county trusts north of the border. The Ulster Trust for Nature Conservation was added in 1978.

The growth of the local movement was strong. During the 1960s, total membership had risen from 3,000 to almost 60,000 but then, during ECY alone, there was another 35 per cent leap. The momentum was maintained throughout the 1970s with total membership passing 100,000 in 1975 and reaching 140,000 in 1982. Between them, the trusts held 500 reserves in 1970, which increased to 1,250 in 1981, covering 42,000 ha. In this year, 1981, there was another change of name when the organisation became the Royal Society for Nature Conservation (RSNC). Today, as the Wildlife Trusts Partnership, it represents forty-seven local trusts and fifty urban wildlife groups. Total membership is estimated at over 250,000 and there are more than 2,000 local reserves which exceed 60,000 ha in area.

The SPNC also became deeply involved with younger people through the WATCH club. In 1971, the *Sunday Times* newspaper promoted a national survey of water pollution. The picture of 10,000 families and young children mobilised to show their interest and concern for the things happening around them encouraged the sponsors to repeat the project with an air pollution survey in the following year. Overwhelmed by the personal commitment of the young participants, which often involved follow-up approaches direct to the polluters, they launched a WATCH club to foster further projects. The *Sunday Times* had hit a sound chord, for the idea flowered and, in 1977, SPNC joined the newspaper as co-sponsor of WATCH, a Limited Trust for Environmental Education. By placing WATCH alongside the county trusts, SPNC in one move invested WATCH with impetus on a local scale and effectively introduced a junior section to the local conservation movement. Within five years, membership reached 15,000 individuals and several hundred affiliated groups.

CoEnCo, too, was keen to extend its influence of young people and set up its Youth Unit in 1978. The WWF already had its Wildlife Youth Service which lectured in schools and, early in 1981, replaced it with an Education Section committed more to teaching the teachers and to supplying them with educational material. Ten years later, aged thirty years, the WWF launched its 'Go Wild' junior section to cater for its new generation. The RSPB, in addition to running its Young Ornithologists' Club, was actively building its educational work. During the year 1983–4, 4,000 teachers attended its courses and a special magazine, *Focus On Birds*, was being distributed to every school in the UK three times a year.

Education was vital to the aims of the conservation movement in the 1970s. Section 13 of the 1975 Act (p. 148) stated: 'A local authority shall take such steps as they consider expedient for bringing the effect of this Act to the attention of the public and in particular school-children'. The decade opened with not one interpretative centre run by the local trusts; by 1982 there were twenty-nine. New technology brought new means of catching and keeping attentions. In 1976, at the Ironbridge Gorge Museum, the CC sponsored the first listening posts for visitors.

Other innovations had particular meaning for the disabled, who began to campaign heavily for accessible reserves and appropriate facilities. The CC published advice to societies and wardens on how to cater for the disabled, and the effects were quickly to be seen; for instance, in the braille plates on the Wildfowl Trust enclosures at Slimbridge.

The growth of the voluntary bodies was not confined to a widening of horizons by the established societies. Increased interest and new areas of concern encouraged yet more newcomers to the scene. One of the most successful was the Woodland Trust which started life in Devon in 1972 and was run for the first five years on an entirely voluntary basis, mainly by its founder, Kenneth Watkins.

> The objects of the Trust are to conserve, restore and re-establish trees and in particular broadleaved trees, plants and all forms of wildlife in the United Kingdom of Great Britain and Northern Ireland and thereby to secure and enhance the enjoyment of the public of the natural environment of those territories.
>
> (Extract from the Trust Deed)

To this end, the trust acquires established woodland by purchase or gift and plants new sites. Its aims are to:

- retain and, where possible, enhance the site as a landscape feature;
- maintain and, where possible, enhance the nature conservation value of the site;
- open the woodland for informal public access (except in very special circumstances or where specifically requested by the donor of a gifted site).

The trust's first purchase was the Avon Valley Woods in South Devon, 40 ha of oak and sweet chestnut under threat of replacement by conifers. New acquisitions followed quickly and, by 1979, it had bought almost 400 ha, while still relying on a membership of under 1,000. The time had come to place the work of the trust on a more substantial basis, and a full-time director was employed with financial help from the CC. Growth became even more rapid. By 1982, membership exceeded 20,000; in the following year, the trust bought its hundredth wood. By 1986, membership was over 50,000. In 1988, its three-hundredth site brought the trust's holdings to 3,700 ha, with £4.5 million spent on woodland conservation. In 1995, 700 sites covered 9,600 ha. Once again, a chord had been struck; the movement was appealing to the hearts and minds of the British people, who value their trees above much else in the countryside. As a result, they rallied round the Woodland Trust which, through its policy of combining conservation with public access, became one of the fastest-growing societies within the movement.

Other initiatives prospered on behalf of our trees. During the 1970s, the most effective was the DoE campaign 'Plant A Tree In '73'. Its aim,

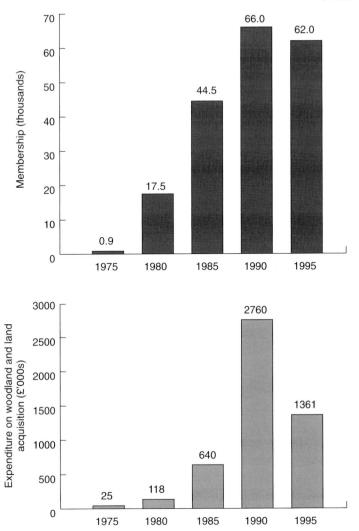

Figure 7.1 Growth of the Woodland Trust

quite simply, was to make an impact upon both the landscape and the public by involving local authorities, amenity groups, companies and individuals in a massive tree-planting promotion. Again, attentions were engaged and the campaign was a success. As a result, the independent Tree Council was formed in 1974 to co-ordinate the work of a whole range of interested parties. It organises the annual National Tree week each November and has done much to further a general appreciation of our

Plate 7.1   Woodlands were not always attractive to the human spirit but, when the conservationists went public in the 1970s, they found there was a niche for a highly successful Woodland Trust. Photograph by Mike Williams. Courtesy of Countryside Commission

trees, not least of all through catchy slogans such as 'Britain Needs Trees', 'Britain's Getting Barer' and 'Plant Now For The Future'.

Some conservationists find fault with such initiatives, claiming them to be nothing more than cosmetic public relations exercises. As such, they should not be underrated, for any increased awareness of or concern for our trees must pay out in the long run. But their fears are wisely voiced, for we ought not to allow a few tree-planting forays to cloud the real picture. Most of our woodland is cosmetic. Britain has only thirty-five native types of tree. Almost half of the ancient 'semi-natural' woodland that remained in 1947 was lost during the thirty-five years that followed. What we have left, and seek to preserve, is a mere remnant.

Between 1950 and 1975, 50 million trees were lost from the British land-scape, 20 million of them through drought, disease and development and the rest through forestry and natural wastage. One third of the few remaining broadleaved woodlands in England and Wales extend to less than 0.5 ha. Between 1978 and 1984, 24,700 ha of broadleaved wood-land was cleared and 11,200 ha underplanted with conifers. Of all the millions of trees planted by the FC between 1965 and 1980, less than 1.5 per cent were broadleaved. Forestry pushed up the wooded area of Britain from 1.1 million ha in 1895 to 2 million ha in 1977 – in other words, from 5 per cent to 9 per cent of the land area – but the average figure for European countries is 21 per cent. Northern Ireland became the barest of all, with a scant 5 per cent of its area wooded – 4 per cent forested upland, 1 per cent on farmland – and much of this declined through age.

So while we can applaud the tree-planting exercises for raising our general awareness of trees, we must never think that they will arrest the continuing decline of our native species. 'Tree-planting is not synonymous with conservation; it is an admission that conservation has failed' (Rackham 1986). No amount of voluntary planting is likely to reclothe our denuded landscape. And only a return to traditional management techniques will revive our existing woodlands as productive resources and rich reservoirs of wildlife.

Another habitat to attract increasing concern in British hearts and minds was the marine environment. Much of what we put on the land and empty into the rivers runs to the sea to combine with what we dump offshore. Yet this very environment that we ply so liberally with sewage, chemicals and other modern wastes, is one of the very richest in terms of wildlife. In the 1970s, it had also been the subject of very little serious study, and the fears of the conservationists lacked the support of scientific proof. But it was clear that many coastal populations were in decline: lobsters and crayfish, scallops and sea-urchins, and many others of which we were only vaguely aware. Over-collecting was frequently given as the cause – the British Sub-Aqua Club alone had 14,000 members – but pollution played an important part. With no machinery to set up official marine nature reserves (MNRs), urgent action was needed.

In official quarters, this came in 1972 in the form of a working party of the Natural Environment Research Council (NERC) which published its findings in the following year in *Marine Wildlife Conservation: An Assessment of a Threat to Marine Wildlife and the Need for Conservation Measures*. Even though the report identified the need for marine reserves, it was to be another fourteen years before the first official one was announced.

Fortunately, the voluntary movement was once again ahead of torpid official policy. By 1972, the Landmark Trust on Lundy Island had declared its environs an unofficial MNR. The same year saw the first moves towards a reserve around Skomer Island NNR off the Pembrokeshire coast. Management plans were drawn up which sought to minimise damage to the habitat and its wildlife by reducing the collection of specimens, disturbance and pollution. The Steering Committee took the proposals to the various interested bodies, and through voluntary co-operation the reserve status largely succeeded. Another 'unofficial' reserve was set up by Torbay Borough Council in Devon, which used the Torbay Harbour Act to designate an underwater reserve in Saltern Cove.

In addition to these initiatives, various campaigns and conferences on the marine environment were set in motion. The UN Food and Agriculture Organisation ran a conference on marine pollution in 1970. The WWF launched 'The Seas Must Live' campaign which sought a global marine conservation strategy that would lead to a more rational and sustainable relationship between man and the marine environment. In Britain, 1977 was declared Underwater Conservation Year. A number of projects and surveys involving the diving fraternity were developed and subsequently expanded into an ongoing Underwater Conservation Programme.

It was from this that another newcomer, the Underwater Conservation Society, was added to the movement in 1977. In its early years, the society remained small; effectively a closed shop for divers and marine biologists. But the need for support from a wider audience resulted in a 1984 revamp which renamed it the Marine Conservation Society (MCS) and gave it charitable status. Early grant support came from the NCC and the WWF, but the MCS was required to become self-sufficient within three years. Early development was faltering and it took a radical review of operations, backed up by not a little nerve and good faith, to bring the society to maturity. The gamble paid off and, during 1985–6, membership almost doubled to stand at 1,446. During the early 1990s, it stabilised at about 5,000 – a small but vital member of the conservation movement with much work to do to drive home to the landbound the international importance of the seashores and shallow seas of Britain. The MCS seeks to safeguard the marine environment through research, education and conservation. It is particularly concerned with pollution, but also investigates the likely effects of planned development at coastal sites, and it was instrumental in having written into the Wildlife and Countryside Act of 1981 a clause to allow for the designation of MNRs (p. 181).

THEORY AND PRACTICE

While many of the new societies were concerned with particular habitats or certain types of wildlife, the principles of conservation as a science in its own right remained largely ignored. The movement, both voluntary and official, could see the need to conserve nature but was not at all clear as to how it should be done. Reserves were being declared all over Britain, some covering thousands of hectares, some consisting of fragile habitats or containing frail communities. How were these to be conserved? What daily steps could be taken in their management to ensure their continuance? Most research had considered the requirements of perhaps only one or two organisms. How could these strands be brought together to relate to the entire environment?

It was in answer to this need that the British Association of Nature Conservationists (BANC) was founded in 1979, with the object of advancing:

- nature conservation and its development;
- public education in, knowledge of and research into nature conservation and its development;
- the understanding and application of nature conservation in the planning of the rural and urban environments;
- the application of ecology and other disciplines to the practical issues of nature conservation.

The BANC publishes a quarterly journal *Ecos* which covers all topics to do with the landscape and environmental politics but, unfortunately, the association has remained a small concern aimed only at professionals in the fields of planning and politics.

While little has been done to advise the volunteer on the mechanics of conservation, much has been achieved to involve the many thousands of people who are willing to dirty their hands and muddy their boots. The leading group in this field is the BTCV (p. 91) which has mobilised a veritable army in getting down to it in the name of conservation. Thousands of work-days per year are spent on practical tasks in the countryside. The media, attracted by stories of schoolchildren and pensioners, businesspeople and mayors, clearing out ponds and picking up litter, has done much to highlight the willingness of volunteers to work on behalf of the cause. Realising that much local potential was still being overlooked, the BTCV published a booklet, *Organising a Local Conservation Corps* in 1972. The NT has been running 'Acorn Workcamps' since the early 1970s, in order to capitalise on its pool of voluntary labour. In 1980, the SPNC estimated that members of the county trusts were giving at least 20,000 days a year to voluntary conservation work. It valued this contribution at £300,000 a year, three times the figure earmarked from their own funds for management work.

It was unfortunate that further encouragement of voluntary involvement was not forthcoming from central government, especially in the light of falling employment and increasing leisure time. At the UN Conference on the Human Environment (p. 137) in 1972, a government paper entitled *Fifty Million Volunteers* called for just such encouragement by the provision of financial support to the voluntary groups so that they might attract maximum public involvement in their work. The paper was effectively ignored. Then in 1978, the NCC began a scheme

Plate 7.2   The BTCV and other groups have mobilised a veritable army of volunteers keen to get down to practical conservation tasks: volunteer workers at Spring Hill, Cannock Chase. Photograph by Mark Boulton. Courtesy of Countryside Commission

of Capacity Grants of £5000 p.a. to each of the local conservation trusts. One of the aims was to increase the activity of volunteers in local management work. Within months, the scheme collapsed through lack of central funding. Not until the introduction of job creation schemes and the Manpower Services Commission did the conservation movement reap any benefit from cheap labour supplied by government, and nobody can pretend that such schemes were introduced on behalf of nature conservation.

Nevertheless, many thousands were employed from the benefit queues. For some small societies, these were their first paid staff, while the NT took on up to 3,000 at any one time. These schemes provided training for many who subsequently took full-time posts in what became one of the fastest-growing fields of employment during the 1970s. Job opportunities in conservation flourished as the voluntary bodies reaped the benefits of increased memberships and local authorities realised the political and recreational advantages to be had from promoting wardened parks and nature reserves. Local authorities, in fact, became some of the biggest employers in conservation and amenity. As cutbacks in job creation schemes were made, however, much of the conservation movement faced financial hardship, redundancy and the failure of many worthwhile projects. The smaller societies were particularly affected in this way. But the idea lives on; in 1995, the Labour party was considering an 'environmental task force' to counter unemployment in the under-25s.

THE NATURE CONSERVANCY COUNCIL

Political inclination to back nature conservation in any convincing way proved to be extremely tardy. Even the official government Nature Conservancy had been swallowed up – royal charter, statutory powers and all – by NERC on 1 June 1965. Although its duties remained the same, its position as a small component of a research body with no public image rendered it impotent as a driving force for conservation. In the four years 1968–71, it spent less than £100,000 on buying reserves, less than half what the county trusts raised voluntarily for land purchase. It was too hard-up to compensate the owners of SSSIs for any conservation measures – with the inevitable result. Many members of staff lost heart and the much-heralded Nature Conservancy hit an all-time low. Official, government-backed conservation in Britain was effectively suspended.

The position became so embarrassing that the government was forced to pass the 1973 Nature Conservancy Council Act in order to restore some credibility to its official conservation agency. Placed under the DoE, the new NCC was responsible for:

- establishing, maintaining and managing NNRs;
- providing advice and disseminating knowledge regarding nature conservation;
- commissioning and supporting research;
- advising ministers on policies regarding nature conservation.

A statement of policies in November 1974 displayed some rejuvenated enthusiasm, with the NCC accepting new obligations:

- towards legislation, particularly under EEC regulations;
- in giving financial support to non-governmental organisations for reserve purchase, staff etc.;
- towards continuing NNR acquisitions and biological survey;
- with the launching of a Geological Conservation Review.

A more tangible product of the rejuvenated NCC was the publication, at last, of *A Nature Conservation Review: The Selection of Biological Sites of National Importance to Nature Conservation in Britain*. This mammoth work in two tomes edited by Derek Ratcliffe finally became available in 1977, although most of the fieldwork for it had been carried out in the latter half of the 1960s. In response to a need which had been identified as early as 1947 by Huxley, the *Review* listed 735 key sites which it was felt could form a useful basis for a series of nature reserves in Britain. These sites, covering 950,000 ha or about 4 per cent of our total land area, were seen as the fundamental ingredients for a wider series of wildlife havens and the conservation of all of them was a minimum requirement. The sites were listed under seven habitat types: coastlands; woodlands; lowland; grassland; heath and scrub; open water; peatlands; upland grassland and heath; artificial ecosystems. The best examples of each type were chosen, taking due account of their size, diversity, naturalness, rarity, fragility, and typicalness, as well as their intrinsic appeal and potential value as a nature conservation site. A factual record of features and a comparative assessment of similar sites governed the decision as to which sites to include. Taking it a step further, the sites were then graded as follows:

Grade 1
- international or national importance (i.e. in Great Britain);
- equivalent to NNRs in value.

The safeguarding of all grade 1 sites is considered essential if there is to be an adequate basis for nature conservation in Britain, in terms of a balanced representation of ecosystems, and inclusion of the most important examples of wildlife or habitat.

Grade 2
- equivalent or slightly inferior to grade 1;
- also of prime importance, but many duplicate features of grade 1 sites which should have priority;
- can be regarded as alternatives to grade 1 sites if absolutely necessary.

Grade 3
- high regional importance;
- equivalent to high-quality SSSIs;
- rating depends on the region in which the site falls (might have attracted a different rating elsewhere).

Grade 4
- lower regional importance;
- still rated a SSSI;
- rating depends on the region in which the site falls (might have attracted a different rating elsewhere).

Many conservationists have seen the grading of sites as 'a tactical blunder of some magnitude' (Adams 1986) but the NCC, treading rather carefully after its recent lull in fortunes, was very aware that the suggestion of a minimum 735 NNRs was likely to be ridiculed in many quarters. Grade 1 and 2 sites combined, for instance, covered almost 1 million ha (Table 7.1).

Table 7.1   Nature Conservation Review Grade 1 and 2 sites

Habitat	Area (ha)	No. of sites
Coasts	263,300	123
Woodland	67,000	230
Lowland grass	65,100	152
Open water	28,000	99
Peatland	63,000	107
Upland grass	427,000	100
Total	913,400	702

The total number of sites is less than the sum of the parts as certain sites contain more than one habitat.

Source: Ratcliffe 1977

To put this in perspective, only 153 NNRs had been designated by 1977. They covered 120,465 ha. Of these, the NCC owned very few; well over a quarter existed by virtue of nature reserve agreements; a similar number were leased; the others were subject to fragmented schemes of management and ownership. The NCC's shopping list was a very tall order; thus the dilution of demands by the grading of sites.

The main strengths of the *Review* were twofold. From a political point of view, it set the test for any future British government wishing to claim an active promotion of nature conservation. Here were the basic landholding requirements of its official conservation body. For the naturalist and conservationist, it was an enviable catalogue of all that was best for Britain's wildlife, right down (or up) to the weevils, with comment on the history of the landscape and the necessary steps for its present-day management.

As a matter of course, the NCC decided to afford SSSI status to each listed site. But designation was a time-consuming task and had not been completed first time round when the Wildlife and Countryside Act of 1981 demanded renotification of all sites. By then, many of the original sites had been lost, mainly due to the practices of forestry and farming. Furthermore, only 20 per cent of them, covering an area of 182,000 ha, were reserves of any sort.

ACTION PLANS AND CONVENTIONS

In fact, in terms of national politics, nature conservation fared poorly in the 1970s. Ministers were still unaware of the growing trend amongst the voters towards a conservationist attitude. But on the international political scene, things were moving more quickly. Unlike national governments, the European politics were not hidebound to the economic creed. Becoming established at a time of increasing environmental awareness, the EEC was just what the movement needed:

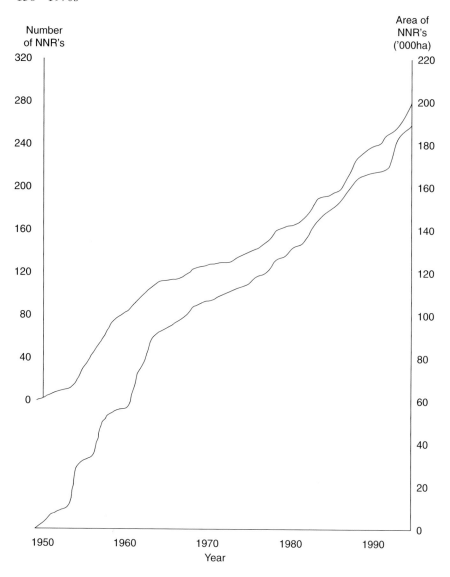

Figure 7.2 Acquisition of the National Nature Reserves (England, Scotland and Wales)

an intelligent audience with no special interest at stake. The Commission, in particular, was staffed by people who had been hired for their intelligence, and because they were salaried and non-political, they didn't have much to lose.

(Rowe 1992)

European Conservation Year, arising from National Nature Week in Britain, became a continental exercise. Two years later, the UN promoted the highly successful Conference on the Human Environment, in Stockholm. Although concerned primarily with the urban scene, it inevitably considered in some depth the effect of the human species across the entire face of the earth. As a result of it, the United Nations Environment Programme (UNEP) was set in motion to deal with international concerns on environment, population and natural resources. The conference called for greater regional co-operation on environmental matters, and consequently EEC leaders met in Paris in October 1972 to consider a common environmental policy. To the delight of the conservationists, they decided that 'improvement in the quality of life and protection of the natural environment are among the fundamental tasks of the Community'.

> In coming to these conclusions, the leaders of the European Community were taking into account the protest by increasing numbers of citizens' groups who rejected the consumer society values that had prevailed in the 1950s and '60s and the economic policies that had sustained this pattern of consumption.
>
> (Ellington and Burke 1981)

The new policies were seen as a way:

> to help bring economic expansion 'into the service of man' by preserving the environment in which he lives and managing the natural resources upon which economic expansion depends.
>
> (*ibid.*)

Notably, the first 'Action Programme on the Environment', launched on 22 November 1973, was aimed primarily at introducing legislation to combat pollution (p. 143). But the second programme, brought in four years later, presented to an economic world a wider spectrum of environmental concerns. This more positive and preventative stance undoubtedly reflects the rapid education of politicians by the conservationists in environmental matters. While sustaining the fight against pollution, the new programme also aimed to achieve:

- international collaboration on cross-boundary problems;
- rational management and protection of environment and resources;
- practical support in fields of research, planning and public awareness, to improve the quality of life.

The programme was based on a number of principles, the most fundamental being:

- the polluter pays;
- prevention is better than cure;
- environmental action must take place at an appropriate level and be compatible with economic and social development;

- environmental considerations must be taken into account in all planning processes.

Amongst its proposals were such avant-garde ideas as ecological mapping and Environmental Impact Assessments (EIAs). The aim of an EIA is to evaluate the total effect of a proposed development on ecological systems and on the quality of life for the affected population. It stands to reason that an EIA must be concluded before any decision on the proposals is made, and the EEC stated its intention to make new developments more transparent to the public and to speed up the planning process. But the implementation of EIAs was very slow in coming (p. 195).

The EC promotes its policies by issuing directives which bind each Member State to obtain certain results, while leaving the State the liberty to decide how such a result should be obtained. Thus, the end is a legal requirement but the means by which it is reached is a matter of choice. Only the member state is accountable, not its individual citizens. Action to ensure compliance by a member state can be taken, either by the EC in the European Court of Justice, or by individuals and associations through the national court of law in the offending country. The Community has introduced numerous measures of environmental importance, but many member states fail to implement them fully. Out of 362 cases of irregularity under review in 1990, Britain was the sixth-worst offender, with thirty-one cases to answer, behind Spain (fifty-seven cases), Belgium (forty-seven), Greece (forty-five), France (forty-one) and Italy (forty). Consequently, a decision was taken in March 1990 to set up the European Environment Agency to act as a watchdog on the application of environmental directives by the member states.

In response to early initiatives within the EEC, and to encourage an environmental background to the formulation and implementation of all EEC policies, a number of voluntary groups got together in December 1974 to form the European Environmental Bureau. The EEB afforded direct access for the voluntary bodies to the EEC and became 'the main voice of citizen opinion on the environment in Europe' (Ellington and Burke 1981). Its stated aims were to:

- promote an equitable and sustainable life style;
- promote the protection and conservation of the environment, and the restoration and better use of human and natural resources, particularly within EEC countries;
- make all necessary information available to members and other organisations likely to assist in the realisation of these aims;
- use educational and other means to increase public awareness of these problems;
- make recommendations in pursuit of the objectives of the bureau and submit these to the appropriate authorities.

As time went on, the role of the EEB became a vital one. At its inception, perhaps ten pieces of European law were environmental in flavour. By 1993, over three hundred could be described thus. The Maastricht treaty even altered the basic aims of the Treaty of Rome from 'an accelerated rising of the standard of living' to 'sustainable and non-inflationary growth responsible to the environment'. It was a concept that every national government struggled to grasp. The EEB can take much of the credit for altering such a fundamental starting point. By working in a bureaucratic fashion with which the politicians were comfortable, it became 'an unofficial working part of the EC ... It was a great environmental coup, performed quietly and with precision' (Rowe 1992).

On a global scale, too, environmental concerns were attracting attention. UNESCO launched its Man and the Biosphere programme in 1971, to investigate problems arising from human uses of the environment. By 1990, it involved ten thousand research workers in 112 countries. A total of 276 biosphere reserves had been designated, where the conservation of biological diversity went hand in hand with the search for models of durable development. The 1970s also saw the ratification of several important international conventions. Their impact and effect varied from one country to the next. Some were uncomfortable thorns in the side of dearly held and unyielding beliefs; others were good, cheap publicity with little about them that was contentious, requiring little action but offering high potential for self-congratulation.

In 1971, the first global treaty on nature conservation was drawn up at Ramsar in Iran. The Convention on Wetlands of International Importance Especially as Waterfowl Habitat has, unsurprisingly, become known simply as the Ramsar Convention. Each signatory state is obliged to consider wetland conservation in its national planning development. By 1994, 657 sites world-wide extended to over 43 million ha. Britain ratified the convention in 1973 and named thirteen sites. Most of these were already protected under some other designation so, with little positive action, Britain was able to congratulate itself on its environmental prowess. Having done so, it then allowed the Ramsar Convention to stagnate until the next decade. Tardily, six more sites were added in 1981 so, as a reminder to the politicians of the potential of the convention, the NCC presented a further 110 areas for consideration. Thus, in 1984, the UK delegation was able to say that 132 sites had been identified as being eligible for designation. This was a fact. It promised only that ten more would be designated 'very shortly' and nearly all would be listed by the end of 1986. By the start of 1994, a total of sixty-two sites had received official recognition. Sadly, designation guarantees nothing for the future. Damaging development can be approved at any time, with only the proviso that an alternative site of similar characteristics be announced in lieu. There is a very real danger that Ramsar will be cheapened by the pointless designation of increasingly poor sites at the expense of better areas' succumbing to development.

Plate 7.3 Marshes and lazy watercourses almost disappeared in the haste to produce more food: Halvergate Marshes, Norfolk, the pilot area which led to the introduction of ESAs and, ultimately, to Countryside Stewardship. Photograph by Richard Denyer. Courtesy of Countryside Commission

The truth is that, for all the pretence, wetlands are still regarded as wastelands, as obstacles to development and progress. Allen (1980) summed up the view: 'With too much water for a picnic and too much land for a swim, wetlands are cherished only by such apparently unnecessary creatures as coots and ornithologists'. Thus it was that, within a year of Ramsar's being ratified, and in spite of the Water Act which required water authorities to consider 'the desirability of conserving flora and fauna', the government was calling upon those authorities in England and Wales to draw up five-year plans to improve drainage. Much central funding via MAFF grants went to encourage drainage schemes, affecting over 100,000 ha of valuable habitat per annum, ten times the figure of the early 1940s. 'The lowland flood meadow, characteristic of much of England and parts of Wales, effectively disappeared within ten years' (Christopher Rose, in Goldsmith and Hildyard 1986).

Animals and plants suffered accordingly. The natterjack toad was lost from its former haunts in southern England and Wales. The fritillary plant was reduced from one hundred in 1930 to no more than a dozen locations. Of the twenty native plants extinct in Britain in 1976, six were plants of the wetland habitat. A further twenty-five wetland species were endangered. Only public outcry could move the politicians and, in 1978 at Amberley Wild Brooks in West Sussex, it did just that. It forced a public inquiry into Southern Water's proposals for a £340,000 drainage scheme affecting 280 ha of rich flood meadow. The plans were quashed on the grounds that the natural communities could not be replaced and that any increase in agricultural productivity would not equal the 'loss' to the public – another hint that the conservationist ethic in the general public was making itself felt with the decision-makers.

Initiatives, such as Ramsar and the Council of Europe's 1976 European Wetlands Campaign, aimed to alert public and politician to the need for a review of the position held by wetlands in the national conscience. The haul is proving to be a long one. In the middle years of the 1980s, 100,000 ha of wetland were still being drained annually. A decade later, 4,000 of our 375,000 remaining ponds were being lost every year. Threats to Britain's estuaries were greater than ever, with 2,000 ha lost annually to industrial development, barrage schemes and recreation, along with more local threats of dumping, commercial exploitation of shellfish, and fish farming. The Cardiff barrage will destroy an entire estuarine SSSI. Such schemes sum up the majority view on the best remedy for a wetland area.

Another convention was the 1973 Washington Convention on International Trade in Endangered Species of Wild Flora and Fauna (CITES). It was ratified in Britain on 3 August 1976, although its recommendations were followed from the beginning of that year. As a result, the Endangered Species (Import and Export) Act was invoked on 3 February 1977. By curtailing the international trade in rare species and their products, it most particularly safeguards threatened species in other parts of the world. It has not been a resounding success, owing to the profits awaiting those who flout the law and the reluctance of officials in some countries to impose the measures strictly.

At the end of the decade, in September 1979, twenty-one states at the European Ministerial Conference on the Environment signed the Berne Convention on the Conservation of European Wildlife and Natural Habitats. It was ratified in Britain on 1 September 1982. Its practical effect was so minimal throughout the continent that the EC had to publish a Fauna, Flora and Habitats Directive in 1988 to pursue the same aims (p. 193). Condemnation of the disinterest shown in the Berne Convention has been widespread. On 1 October 1985, the UK became party to the Bonn Convention on the Conservation of Migratory Species of Wild Animals; this too has had little practical effect in Britain.

In 1977, the ICBP had set up a Migratory Birds Committee and launched an appeal to support a new initiative to reduce the wholesale slaughter of migrating species. It was estimated that 150 million birds were being killed every year in Italy; that between 5 million and 10 million were being trapped or shot in south-west France; that, in Malta alone, 1,500 birds of prey and a quarter of a million turtle doves were dying. In the light of such slaughter, all steps taken at home to preserve these birds appeared to be a waste of time. By April 1979, the EEC was sufficiently moved to issue its first directive not concerned with pollution. This Directive on the Conservation of Wild Birds aimed not only to curtail the mass killings but also to enhance habitat conservation. As with so many directives and conventions, however, it was not readily accepted by those nations most affected by its contents. It introduced yet another designation: that of Special Protection Area (SPA). In 1986, Britain claimed to be considering 149 sites for such designation and the NCC set about ensuring that all were notified as SSSIs. In 1990, the NCC denounced the rate at which SPAs were being announced and presented a revised list of 210 worthy sites. In the ensuing flurry of embarrassed political activity, seven more were designated in Scotland, bringing the UK total to a mere forty-six.

Of the EEC, Ellington and Burke (1981) wrote:

> It was the growing number of consumer protection groups, nature conservationists and environmental pressure groups that sprang up in the late 1960s and early 1970s that, to a large degree, caused governments throughout the Community to establish new ministries to deal with environmental problems.

Likewise, at home, any moves by the politicians towards conservation were as a result of the efficacy of the voluntary movement. ECY had mobilised a sometimes vociferous, always genuine mass of concerned voters who, as the decade progressed, came to be afforded a little more than lip-service by their Parliament.

DISEASE AND POLLUTION

The difference between the 1960s and 1970s is clear. The earlier decade was a time of overall optimism; of flower-power, hippie happiness and generally relaxed well-being. The purveyors of Doomsday despondency were heard in some quarters but their overstated claims could not penetrate the aura of optimism. They were just another fringe body of a colourful decade. But, during the 1970s, fundamental changes began to occur. Certain Doomsday forecasts were acted out. A feeling of unease undermined the satisfaction of the 1960s and in crept a fear of how far the Doomsday claims might prove to be true. When oil prices rocketed and supplies were threatened in 1973, the instability of a product upon which we had come to depend so heavily chilled the ardour of economic common people.

Dutch elm disease took a virulent hold after half a century of quiet brooding in British woodlands when a new strain entered from North America. It spelt disaster both for the landscape and for the morale of its guardians. By 1972, 11 per cent of elms in the Midlands and the south-east were dead or dying. All treatments failed and the experts and conservationists could only suggest felling the stricken trees, not realising that this course of action would prove as costly and futile as had treating the *Torrey Canyon* birds with detergent. By 1978, half of Britain's 25 million elms were affected and the disease was being left to take its course which, in the long run, proved to be the best policy. Nature's own fail-safe mechanisms seem, for the time being, to have halted the destruction with some assumed immunity. But long after it was dropped from the newspaper headlines, the disease continued to devastate the landscape and, in 1982, MacEwen and MacEwan wrote, 'It seems improbable that new broadleaved planting is even keeping pace with losses from Dutch elm disease, let alone stemming the insidious decline in existing wood-lands.' Of all the fundamental changes pervading the air of the 1970s, a basic fault in the economic machine, together with a landscape stripped of trees, seemed to matter most.

Pollution – 'materials in the wrong place' (Durrell 1986) – continued to attract attention.

> Pollution is a threat of such immense and sinister proportions that the average citizen may be forgiven for thinking that there is nothing much he can do about it as an individual.
>
> (Gresswell 1971)

The EEC's First Action Programme for the Environment was concerned solely with pollution. It aimed to:

- define quality objectives for fresh and sea water;
- protect the aquatic environment against pollution;
- survey and monitor water quality;
- control pollution by certain industries;
- undertake the requirements of international agreements, etc.

At home, the Royal Commission on Environmental Pollution was set up in 1970. To its credit, its first report was quickly made available in February 1971. It commented on existing problems and likely trends and called for action, such as a reduction in the dumping of toxic wastes, a close appraisal of coastal pollution, better facilities for water supply and treatment of sewage, improved methods of monitoring pollution and better training in pollution control. It did not, however, add its voice to the clamour for banning persistent pesticides, recommending only an ongoing appraisal of the situation. While stating that nobody had a right to pollute the environ-ment without control, it nonetheless pointed out that anti-pollution requirements must be made upon known facts and must take into account the cost-effectiveness of proposed schemes and the willingness or other-wise of the public to pay for them. In 1974, responsibility for water quality in England and Wales passed from the local authorities to regional water boards, but these never really shook off the aura of local council and they achieved very little in tackling pollution.

Surveys of river quality at five year intervals began in 1970. The findings were depressing, although there was some improvement during the first decade. Then the trend was reversed and overall quality has been steadily falling since 1985. Coincidentally, it was not until 1985 that information on discharges into our rivers was made public and, perhaps in order to soften the blow, government chose that moment to alter the way in which discharges were shown, by replacing absolute limits with percentile limits. Kinnersley (1994) describes this statistical obfuscation as 'a sort of environmental betrayal'.

The year 1974 saw 'one of the more far-sighted pieces of legislation of the past decade' (Dawson 1980). The Control of Pollution Act required the regional water authorities in England and Wales and the river purification boards in Scotland to control the pollution of our rivers. So thorough was the Act that the EEC based its own Waste Disposal Directive upon it. The shame, however, is that whole chunks of it were not implemented before it was overtaken more than a decade later by European directives. This situation is officially blamed on a shortage of finance, and Johnson (1983) describes it as 'a stain on Britain's record of environmental management'.

Another innate weakness, perhaps caused by shortage of cash, was the unwillingness of most water authorities to pursue their duties enthusiastically. The Yorkshire Water Authority was one of the few to have felt the full force of public outcry for this failure. Some 476 km of polluted river under its 'care' was still deteriorating in quality at a time when trends elsewhere in Britain were towards greater cleanliness. Over 70 per cent of effluent emptied into these stretches was untreated. In three years, the Authority had taken out only eight prosecutions against the offenders. The *Yorkshire Post* ran a series of articles naming the worst culprits and highlighting the authority's pittance of a budget. Feelings ran high; many local people, including schoolchildren, carried out surveys and monitored pollution levels. Local MPs were inundated with complaints. Within weeks, the authority had revamped its pollution prevention department with new laboratories and a 250 per cent increase in budget. In the next two years, it took out thirty-five prosecutions, while the polluters themselves did the rest. All those named in the articles took positive steps to clean up their act.

Not all companies have found that preventing pollution costs money. When North British Distilleries were legally required to stop discharging their residue into local rivers, they took to drying it and marketing it as cattle feed – and made 100 per cent return on investment in less than a year. The 3M company reviewed its operations to reduce waste. In the UK, it found fifty ways of doing so and saved £2.5 million. Worldwide, these steps saved the company £235 million over twenty years!

In spite of the improvements of previous decades, air pollution continued to be a source of worry in the 1970s, but the effects of lead,

Plate 7.4   Pollution and industrial blight caused a sadness that found relief in new ventures to clean up our cities. Companies indulged in some cheap sponsorship but, when they started to seek cleaner ways to behave, some also reaped a financial reward. Photograph by Martin Page. Courtesy of Countryside Commission

dust, smoke and other air pollutants remained largely unmeasured and misunderstood. Not until the acid rain scare of the 1980s did this form of pollution become a front-runner for the conservation movement (p. 186).

SEALS, OTTERS, BATS AND BADGERS

If pollution seemed too technical, too scientific, or indeed too 'immense and sinister' for the armchair conservationist, there were other signs of disruption to whet the appetite for the cause. Seals and otters are always close to the hearts even of those whose eyes have never seen them.

In 1963, a world total of 46,000 grey seals was estimated, with 78 per cent of them around British coasts, and 25 per cent in Orkney alone. The population was found to be doubling every nine years. Not only did the seals begin to deplete fish stocks, but they also drove out ground-nesting birds by sheer weight of numbers and deprived themselves of food and space. It was also clear that the so-called common seal was in rapid decline. So a yearly cull of grey seals on Orkney was authorised by the Department of Agriculture and Fisheries in Scotland (DAFS). In answer to the need for careful balancing of seal numbers, the 1970 Conservation of Seals Act was introduced. The close season on grey seals remained the same as under the 1932 Act (p. 47), from September to December and, for the first time, the common seal was afforded protection with a three-month close season from June to August.

The annual culling went on into the 1970s, a valid conservation measure on the face of it and one which cannot have threatened the grey seal since, by 1977, the programme was stepped up in the face of ever-increasing populations. But in 1978, public opposition rallied. A total of 5,800 defenceless cuddly seals were to be slaughtered in Britain by Norwegian gunmen. Greenpeace gave the matter such high media profile that DAFS had to withdraw permission in answer to the general outcry. Subsequent culls have regularly been banned because of public concern. The scientific integrity of this result is open to question but popular conservation, based on sentiment, had nevertheless won itself another victory. In 1992, Britain hosted 100,000 grey and common seals – over 40 per cent of the world population – and ugly scenes of unofficial culling were to hit the beaches of Orkney.

For the otter, there can be no question. By the start of the 1970s, populations in England and Wales were negligible due to disturbance, pollution and habitat loss. Hunting had practically ceased through necessity although, in Ireland, the demand for pelts remained great enough to threaten the remnant population. It seemed likely that the otter would become extinct in most of Britain within a very few years. As a result, the Joint Otter Group was formed in September 1976 by the NCC, the SPNC, the Institute of Terrestial Ecology and the Mammal Society. The last organisation was running a survey from 1973 to 1979 and the Joint Otter Group set up a more detailed survey in its last two years, for which the author was a field worker in Wales. The Group instigated the novel idea of otter havens, of which there were two by 1978, and it took steps to press for urgent legislation to protect the species.

Two options presented themselves: either a new and specific Otter Act or an addition of the otter to the list of protected species under the Conservation of Wild Creatures and Wild Plants Act (p. 149). The former would have been slow to reach the statute books if indeed it could overcome the likely opposition; the latter, while offering an easier option, would possibly prove less effective in the long run. After much discussion involving the hunting lobby and other interests, the otter was added to the protected list of the 1975 Act in January 1978. But the Act only protected species that were already rare and, in the case of the otter, only in England and Wales. The legislation could not ban but merely limit hunting, which had been a major factor in the animal's decline through the disturbance it caused. The hunting lobby was likely to fight for its removal from the protected list as soon as populations showed any hint of rising, which they subsequently did in some areas, giving the UK a total population of about

7,000 in 1995. It was a Catch 22 situation, unsatisfactory in many ways, but better, for the time being, than no protection at all.

Bats were not so well loved, but thanks again to a Mammal Society survey, this time in 1971, those in conservation circles began to hear of serious declines. Seventeen species of bat occur in Britain; fourteen have been known to breed. The mouse-eared bat, which only colonised Britain in the 1950s, was not found after early 1991, so the rarest now are the greater horseshoe and the Bechstein's bat. Apart from their bad publicity based on spurious superstition which does not win them many champions, bats are also in danger from destruction and disturbance of roosting and breeding sites in caves and tunnels, buildings and trees, and from pesticide concentrations in their insect food. Early steps to help them revolved around locking grilles at the entrance to a favourite site and the provision of bat boxes. Media interest was aroused by the thought of helping such unattractive creatures but, by their very nature, bats still fail to excite sufficient support (p. 196).

Public opinion was more willing to rally on behalf of the badger. Numbers had been declining since 1945 but very little concern was evident until the animal made the news in 1971. In that year, one unfortunate individual was found to be carrying bovine tuberculosis, a disease which MAFF had been trying to wipe out since 1935. Further evidence showed beyond doubt that badgers could transmit the disease to cattle. As discussions began on methods of controlling badgers, the ancient and continuing 'sport' of badger digging came under surprised public scrutiny. Most people were astonished to hear how common and widespread was the practice of dragging badgers from their setts with metal tongs, mutilating them and then pitting them to death against terriers.

The Badgers Act 1973 enjoyed a relatively easy passage onto the statute books. It became illegal to:

- wilfully kill, injure or take a badger;
- cruelly ill-treat a badger or to use badger tongs;
- dig for, or to mark a badger;
- offer for sale or possess a live badger;
- use certain firearms to kill or take a badger.

However, an owner or occupier of land, or any authorised person, could kill or take badgers just as before, so the Act had limited effect. A clause that allowed the Secretary of State to declare areas of special protection, in which even landowners had to show good reason for killing a badger, was almost irrelevant, since by 1980 only one such area existed.

The fact was that, notwithstanding the Act, MAFF was pressing hard for powers to cull badgers as part of its bovine tuberculosis control. That power was granted in August 1975, when a clause in the Conservation of Wild Creatures and Wild Plants Act amended the Badger Act and allowed widespread gassing by the ministry. In the next thirty months, 1,000 setts were gassed in Gloucestershire alone and the 'treatment' was extended to parts of the country where bovine tuberculosis did not exist. When gassing was belatedly declared inhuman, trapping and shooting replaced it.

In April 1986, the NCC published its report, *Badgers and Bovine Tuberculosis*, commonly referred to as the Dunnet Review. With it, official policy moved from pursuing total eradication of the disease in badgers to controlling its transmission to cattle. The killing of badgers was restricted to farms where the disease was confirmed, and a 50 per cent reduction in the number of animals culled was expected.

Seventeen years of culling had cost MAFF £11 million by 1993 and had failed to reduce tuberculosis in cattle. In spite of the 1992 Badgers Act, which reiterated and theoretically strengthened many of the sentiments of its predecessor, a new test was devised which relied upon the removal of individual badgers to captivity for up to three days, regardless of any dependent cubs or other humane considerations.

The 1992 Act, in fact, banned all unlicensed disturbance of setts. There were thought to be 250,000 animals in the UK, in spite of a yearly total of 47,000 road casualties. They inhabited about 90,000 setts, of which over 10 per cent were dug out and almost 16 per cent were blocked up during a three year period. For all the legislation, groups such as the Mammal Society still had work to do to achieve any real protection for the badger.

It was, however, ably assisted by a number of small, localised groups which began to rise to the defence of the animal. It is an interesting phenomenon which owes more to the local community scene than it does to the mainstream of the conservation movement. These groups consisted of neighbourhood watchmen concerned to see 'their own' badgers left undisturbed. Their horizons originally stretched no further than this. However, by January 1986, there was a need for the National Federation of Badger Groups, which was formed to co-ordinate the work of the ever-increasing local groups.

THE CONSERVATION OF WILD CREATURES AND WILD PLANTS ACT

While legislation to protect certain animals and birds remained disjointed, confusing and in many cases unworkable, protection of our plant life remained non-existent. Over a dozen species were lost to Britain between 1900 and 1970, half of them in the final twenty years. In 1977, two publications brought the problem into sharp focus. The SPNC published its *Red Data Book on Vascular Plants*, which listed 321 species found in fifteen or fewer 10 km squares. The number of plants in this category had probably doubled in the preceding fifty years. Each species was given a 'threat number' based not only upon its rarity, but also on its rate of decline, its attractiveness and its accessibility. Starfruit, a water-plantain once found in southern ponds and ditches, won the dubious honour of

Britain's rarest plant. It had been recorded once, in Surrey, since the start of the decade, its decline largely due to the wholesale drainage of its sites. Careful management of habitat has since eased its situation. Corncockle, also listed amongst the eight rarest plants, and now extinct, had appeared on a very different blacklist of agricultural weeds in the sixteenth century.

Ratcliffe (1977) identified 1,700 species, of which 280 were regarded as rarities, having been found in fifteen or fewer 10 km squares since 1950. Of the 406 species dependent upon man-made habitats or activities, forty-five were viewed as rarities. Both publications made mention of other species declining rapidly towards the endangered status.

Meanwhile, all efforts to procure some legal protection for plants continued to be unsuccessful. A Wild Plants Protection Bill of 1967 had failed through lack of time. Another attempt in 1974 met with the same fate. Then, late in 1974, Peter Hardy MP won the right to promote, as a Private Member's Bill, a revised Wild Plants Protection Bill together with a Wild Creatures Bill. The combined Bill enjoyed an unopposed second reading in the House of Commons on 24 January 1975 and became law as the Conservation of Wild Creatures and Wild Plants Act 1975. As far as creatures were concerned, it effectively gave the same protection to scheduled species as did the Badgers Act (p. 147). All wild plants were protected from uprooting by anyone other than an authorised person, and

Table 7.2 Plants protected under the Conservation of Wild Creatures and Wild Plants Act 1975

Species	No. of locations (1975)
Oblong woodsia (*Woodsia ilvensis*)	4 sites
Soldier orchid (*Orchis militaris*)	2 sites
Blue heath (*Phyllodoce caerulea*)	1 area
Alpine sow thistle (*Cicerbita alpina*)	1 area
Monkey orchid (*Orchis simia*)	3 sites
Spiked speedwell (*Veronica spicata*)	1 area
Alpine woodsia (*Woodsia alpina*)	15 sites
Alpine gentian (*Gentiana nivalis*)	
Spring gentian (*Gentiana verna*)	1 area
Teesdale sandwort (*Minuartia stricta*)	1 area
Ghost orchid (*Epipogium aphyllum*)	
Snowdon lily (*Lloydia serotina*)	1 area
Wild gladiolus (*Gladiolus illyricus*)	
Mezereon (*Daphne mezereum*)	
Cheddar pink (*Dianthus gratianopolitanus*)	1 site
Drooping saxifrage (*Saxifraga cernua*)	3 sites
Lady's slipper (*Cypripedium calceolus*)	1 site
Red helleborine (*Cephalanthera rubra*)	
Killarney fern (*Trichomanes speciosum*)	
Tufted saxifrage (*Saxifraga cespitosa*)	
Diapensia (*Diapensia lapponica*)	1 site

certain scheduled rarities were further protected from being picked, uprooted or otherwise destroyed unless as an unavoidable result of 'good agricultural or forestry practice'. The Act called upon the NCC to advise the Secretary of State every five years, or at any other time, so that the schedules of species might be updated either by addition or by removal of names.

This Act can be heralded as a first attempt to provide a framework within which all types of wildlife might receive protection. As such, it represented a large step forward for nature conservation. But in reality, it was crippled by severe limitations. The small number of species scheduled reflected just how very scarce they had to be before attracting protection. In many cases, such rarity meant that the battle had already been lost; that protection had come too late. Furthermore, the hasty removal of a name from the schedule, perhaps as a result of a perceived recovery, might ironically spell an end to that species. The conservation movement itself had long since moved on from the philosophy of protecting only the rare, but this piece of legislation clung to its worn-out precepts. Mabey (1980) states: 'I do not believe that this is an expression of the way we, as a society, regard our wildlife heritage at present'. But can we be so sure of that? It is hard for the ardent conservationist to believe that others do not share his foresight and concern.

The facts speak for themselves. Even in the mid-1970s, this was as far as we could go towards comprehensive wildlife legislation. The conservationists were not a sufficient thorn in the side of the politicians to warrant any more than this type of treatment. They had not attracted enough voters to the cause. So with the situation looking grim for the handful of creatures and plants that were scheduled, the conservationists were happy to accept the little they were given. Meanwhile, the politicians made much of their efforts on behalf of our wildlife, and Mabey pierces their facade with precision. Of the schedule of protected plants, he writes:

> They are not the rarest plants in Britain nor ... the most threatened. Nor ... are they the most beautiful, popular or widely picked. If one was looking for a single quality possessed by all twenty-one it might be 'glamour' – which perhaps best suggests their combination of scarcity, reputation and allure. If this was a deliberate consideration, it was politically very astute, recognizing as it does that in such difficult areas *laws function more as propaganda* than as checks on specific actions.
>
> (*ibid.* my italics)

An overriding worry for the conservation movement must be the danger of rarity becoming an attractive quality.

> a county loses its otters, a village its cowslips, a farm its swallows ... they are the ways that extinction begins, and what, at grass-roots level, the conservation movement should be fighting against. If there is a single

starting point, it must surely be in attempting to intervene in the process by which what is common becomes scarce, to prevent the stage of rarity and imminent extinction being reached.

(*ibid.*)

It is very easily done, Mabey says,

to make the process of adjustment, to trim this year's expectations to last year's realities, and to forget altogether about those sites where there can no longer be any expectations.

(*ibid.*)

The conservation movement of the 1970s still faced an uphill struggle to change attitudes. For all its 'going public' and winning new followers, too many vested interests and too much inertia were working against the dream's coming true. There was a very real danger that the movement would sit back and rest on its collective laurels, having mistaken its new audience for an active following in official circles, when official policies were still restricted to the old, well-trodden paths.

The majority of British people – and this was the crunch for the thriving conservation movement – were still intent on pursuing economic growth through antiquated policies that were condemning the environment. They would listen to the arguments, shake their heads at the inevitability of it all, and completely fail to recognise the real dangers of their chosen course. They never thought there was any real alternative.

FARMING FOUND OUT

Fundamental attitudes needed revision, and those surrounding farming and forestry were some of the most urgent and the most entrenched. The Agriculture Act of 1947 begat an industry that could provide vast amounts of produce for home use and export. It had deservedly become a wealthy industry and its wealth had brought it power and influence. Vested interests here were greater than most.

The severest threat to our wildlife is undoubtedly loss of habitat. Agriculture, practised on 80 per cent of Britain's land surface, has always been the creator of the landscape. Let us consider just one detail of the agricultural scene; the hedgerow. It is estimated that, in the mid-1940s, Britain could boast 980,000 km of hedgerow, over 80 per cent of it in England and Wales. If these hedges averaged two metres in width, they would have amounted to almost 200,000 ha. In the forty years following the Agriculture Act, 300,000 km were grubbed out in England and Wales alone, 85 per cent of them as a result of agriculture (Table 7.3). Vastly more was spoilt through lack of management. Eastern England saw the greatest losses between 1962 and 1966 when an incredible 5,600 km a year were being removed. By 1977, the region was stripped to no more than

Table 7.3   Rate of hedgerow removal (England and Wales) 1945–93

Period	km p.a.
1945–60	7,200
1960–66	16,000
1966–70	7,200
1970–74	4,800
1974–78	3,200
1978–85	6,400
1985–88	5,000
1988–93	3,600

Far greater losses have been suffered through lack of correct management: over four times as much (14,500 km) annually in 1990–3 (ITE 1994).

8–13 km of hedgerow in each kilometre square, compared with 24–29 km in the west and up to 40 km in particularly favoured areas (Ratcliffe 1977).

The importance of hedgerows lies in the diversity of their qualities. Many are remnants of the great primeval forest. Farming introduced others to limit the movement of livestock and of pest species, to provide shelter and to reduce soil erosion. Their value in each of these aspects is open to discussion, but, for many landowners, the advantages remain relevant. Similarly, while few species of animal or plant depend wholly upon hedgerows, many do colonise them or look to them for shelter and a source of food. Between 1978 and 1993, the diversity of species in our meadows and woodlands fell by 14 per cent. As broadleaved woodland, copses and rough field corners are cut and cleared, the importance of hedgerows continues to grow. It is estimated that 10 million birds of sixty-five species use the hedgerows as nesting sites. Twenty-three types of butterfly and twenty-one mammals breed in them. Some 250 species of plant colonise them regularly and, because other habitat is threatened too, up to thirty of them could become extinct in Britain if hedgerow removal continues apace. Many of the plants are fugitive from the fields they once adorned, hiding in the hedgerows from the machinery of modern agriculture and the worst effects of pesticides. Today's greatly diminished network of hedgerows, estimated at 675,000 km or 135,000 ha, still approximates the total extent of our NNRs and represents a very valuable wildlife resource.

Then there are the other considerations, both spiritual and recreational. Why did we single out the hedgerow from the whole farming scene? Why, if it is so much diminished and further threatened, does it still portray such a vital part of the rural landscape in British hearts and minds?

Habitat and species loss may be the concern of the natural historian worried about the fate of less common species which do not adapt to habitat alteration. But for the majority of us, it represents something

Plate 7.5 Hedgerows were lost through neglect and grant-aided removal, but was it really in the farmers' interest to leave great swathes of exposed and unvaried land? Photograph by Peter Hamilton. Courtesy of Countryside Commission

else. The still unfamiliar prairie-like sweeps of land in parts of lowland Britain, divided only by bare-edged roads and drainage ditches, jar the senses as they destroy the recreational value of old footpaths and bridleways.

(Johnson 1983)

The human spirit needs the hedgerows; yet economic attitudes have become ingrained. How, in this day and age, do you make this argument stand up against an economic coup? For the conservationist, there are two hopes: to prove the false economy to the farmers and to bring public pressure to bear upon the politicians.

As early as the 1970s, the wisdom of wholesale hedge removal was being questioned in the United States. It was becoming clear that enclosures of more than 40 ha were no more economical than the smallest of fields.

Currently some farmers are grubbing up hedges, in a pathetic attempt to imitate the methods of America, and the British Ministry of Agriculture, never very smart, is even encouraging this by giving grants for the purpose. It is reckoned that hedges are going at 10,000 miles [16,000 km] a year.

(Taylor 1970)

The pressure brought to bear upon the politicians resulted in the withdrawal of MAFF subsidies in 1972. This altered the complexion of the economic argument, and the rate of hedge removal temporarily decreased. But the politicians would go no further than that. They did not heed calls from the CPRE for grants to preserve hedges or for planning powers to prevent unwarranted removal. The only hope lay in voluntary codes of practice, but these have not been equal to the task and hedgerow removal increased again as the economic balance began to tip back in the 1980s.

Hedgerows are just one component of the landscape moulded by agriculture. Eighty-five per cent of those lost were uprooted on its behalf. The NCC attributes other stunning losses to the same industry. Ninety-five per cent of lowland grasslands and hay meadows disappeared or were 'improved' between 1947 and 1983 and much of what remained was severely damaged. Eighty per cent of chalk downland went, along with much characteristic flora which consists, in good habitat, of 120 species of flowering plant and twenty types of grass. During the 1950s and 1960s, a quarter of Dorset's chalk downland went under the plough, finally reducing the early nineteenth-century total of 28,000 ha to a mere 3,000 ha. Wiltshire lost half its downland between 1935 and 1980. Even so, three-quarters of Britain's remaining downland is found in these two counties and in neighbouring Hampshire, thanks mainly to the presence of the military there. Even though MAFF grants for lowland ploughing were withdrawn in 1972, agriculture continues to pose the greatest threat to this habitat through the application of pesticides.

Over 40 per cent of lowland heaths had been 'improved' for agriculture, or forested, with consequent loss of wildlife. The famous Breckland heaths of East Anglia were reduced from 22,000 ha in 1880 to just 7,500 ha a century later. Fifty per cent of our fens and mires were drained beyond recognition and worth, and again East Anglia suffered heavily. Of almost 340,000 ha of natural fenland in the region, all but 10,000 ha had been lost before 1934, leaving less than 1,000 ha. Throughout England and Wales, natural lowland woodlands were lost to forestry and farming. Since 1947, eight counties had lost over half their total. Gwent suffered a 67 per cent reduction. Nineteen species of wild flower were lost in the first eighty-five years of the century. Kent, Northumberland and Staffordshire were amongst the counties that had lost a tenth of their species since 1950 alone. Lincolnshire lost fifty plants during this time. Nationally, 20 per cent of wild flower species were threatened.

The NCC concluded that 80 per cent of bird species and 95 per cent of butterflies would disappear if all farmland were fully 'improved'. Twenty-four types of butterfly use natural pasture, none uses reseeded ryegrass. Few species are so specialised as to depend upon only one or two types of plant, 'so that the extinction of a rare arable weed may be of less overall significance to insects than the reduction in population size of a group of related plants which remain relatively common' (Ratcliffe 1977). Almost two hundred plants are 'weeds' of arable land; half of them are uncommon and decreasing. Swine's succory and thorow-wax became extinct in 1970; interrupted brome in 1972. This signals serious danger to many of our insects and the creatures which in turn depend upon them. The SPNR (1971) was laconically matter-of-fact about it:

> there is clearly no future in lowland England for floristically rich meadows or pastures. These will survive eventually only where County Trusts and other bodies have been able to acquire and manage them on traditional lines.

The NCC charted these changes in the British landscape from a starting date of 1947, the year of the Agriculture Act and, for that matter, of the Town and Country Planning Act. As contemporary calls mount to bring farming and forestry within the sphere of planning, it becomes more difficult to remember that, in 1947, farming was seen as the guardian of the countryside, with its health and well-being of primary concern. But the business that was ushered in by the new Act was a very different business indeed. Money begets money, and money begets greed. The agriculture machine rolled on, devouring much in its path to feed the people of Britain, either directly, or indirectly through exports. And a very successful machine it was in producing the results it was set up to produce. Very little was found to fault it. There was the brief flurry of conservationist activity over pesticides in the early 1960s, but that soon passed. Public opinion went back to sleep; the farmers were doing their job. The repose was not shaken until the publication by the NCC of those figures of loss in 1984. During the 1970s, the conservationists failed in their duty to alert the wider public. Moreover, they were only just beginning to talk to the farmers. Harrison (1973) points out that the valuable work of Dr Janet Kear on wildfowl and agriculture was well known to naturalists. 'Unfortunately, virtually nothing on the subject has appeared in the popular farming press, so that the results are quite unknown to the average farmer.' In these circumstances, the farmers might be forgiven for pursuing their aims with little regard for wildlife; those blessed with an understanding of it – the conservationists – were keeping it to themselves.

FARMS, FORESTS AND PARKS

The most notable conservationist approach to the farmers, and one that is thriving today, was the formation of the Farming and Wildlife Advisory Group (FWAG). This highly successful organisation arose from an exercise run by the RSPB in July 1969. One hundred farmers, farming advisors and conservationists got their heads together at a practical conference held on a farm near Silsoe in Bedfordshire. They agreed to set up FWAG to identify the problems in reconciling the needs of farming with those of landscape and nature conservation; to find areas of compromise and then to spread the word. Similar exercises followed in other parts of the country. Local FWAGs were formed and, early in 1971, the SPNR and the RSPB jointly financed the appointment of a national advisor. The movement steadily gained momentum and, by 1985, there were fifty FWAGs in England and Wales and fifteen Farming, Forestry and Wildlife Advisory Groups (FFWAGs) in Scotland. Most of their work is performed voluntarily, although many of them have a full time advisor. The FWAGs have achieved a great deal by galvanising at least the sympathetic landowners to take action on behalf of wildlife, while also awakening many conservationists to the real needs of the farmers. It was a tentative bridge between the two sides which has heroically stood the test of time.

During the 1970s, the CC was very concerned at the rate of moorland reclamation, particularly in the Exmoor National Park. NCC figures showed a 30 per cent national loss of this habitat between 1947 and 1983 and attributed a quarter of it (150,000 ha) to reclamation for farming. Twenty per cent of Exmoor's 23,600 ha became farmland between 1947 and 1978. Figures for the other National Parks are little better, but, with MAFF grants far outstripping the resources of the park committees, the situation looked set to go on. In desperation, the CC reported the committees to the Secretary of State in 1977 for failing in their duties. Lord Porchester was called upon to review the matter and suggested that the committees be empowered to make Moorland Conservation Orders. Landowners of affected areas would be prevented from reclaiming them and would receive a one-off payment as compensation. He also called upon the government to pay 'conservation grants' towards the cost of conservation in the National Parks. These suggestions were embodied in a Countryside Bill of December 1978, but met with severe and successful opposition from farmers who resented the restrictions on their freedom to follow their own policies.

The Bill fell, but not the idea of paying land-owners to conserve wildlife. The NCC, not without much heart-searching for alternatives, had to admit that nature conservation did not pay in financial terms, least of all for farmers, and that any attempts to suggest to them a duty towards wildlife would be spurned in an economic world.

The year 1978 saw the first official recognition of the conflict in the Strutt Report, *Agriculture and the Countryside*. This called on authorities to set up consultative groups involving the various parties: MAFF, the National Farmers' Union, the NCC, the CC, the FC, tourist boards and, interestingly, the general public. The problem had become so great, and was now so much in the public mind, that something should be seen to be done. The stage was set for action towards understanding and compromise in the 1980s.

Forestry interests had to be involved, too. On the moorlands, forestry was a bigger threat than farming. Most of the 1 million ha planted between 1950 and 1980 had been on moorland. Planting would continue at 20,000 ha a year. Much of our native broadleaved woodland was being replaced by conifers. Ironically, in the early 1970s, farmland itself was disappearing under trees at the rate of 20,000 ha a year. Nearly 8 per cent of Britain, 1.7 million ha, was wooded. Private woods covered 800,000 ha. The FC was the biggest landowner in Britain, owning 1.1 million ha, of which two-thirds was wooded.

The FC had, at least, played host to recreation and conservation interests quite early in its history. After its appointment of a landscape architect in 1964, it created a Conservation and Recreation Branch in 1970 and began to train its own wildlife officers. In 1972, it published *Wildlife Conservation In Woodlands* by Richard Steele of the NC. By 1975, there were thirteen Forest Nature Reserves amounting to 960 ha, most of them owned by the commission.

To cater for the amenity interests, there were seven Forest Parks by 1971, covering 200,000 ha. These were: Argyll, Glen Trool, Glen More and Queen Elizabeth in Scotland; the Forest of Dean and Wye Valley Woods in England; Snowdonia in Wales, and the Border Forest Park between Scotland and England. But, strange as it may seem in retrospect, forests were not a popular haunt for recreation in Britain, and the review of FC policy in 1972 called for further energetic development of the amenity side of its work.

While forestry concerns were striving for a better press on amenity matters, new moves in wider countryside recreation seemed to stagnate. The country parks and picnic sites multiplied in number, but further new initiatives were not forthcoming. The activists, perhaps, were taking breath after the introduction of the Countryside Acts. But in a decade of increased leisure time, not least of all due to unemployment, more positive steps towards a wider appreciation of our countryside might have been expected.

The National Parks were going through a spell of heart-searching. The Policies Review Committee was drawn up in 1971:

> to review how far the national parks have fulfilled their purpose for which they were established, to consider the implications of the changes that have occurred and may be expected in social and economic conditions and to make recommendations as regards future policies.

In its 1974 Sandford Report, the committee aired its expectations of seeing visitor numbers and demands for facilities double by the year 2000. It commented on a variety of problems that were worrying the Parks administration, but it did so without urgency and enthusiasm and had little practical effect as a result (p. 225). Policies, reviewed, remained largely unchanged.

North of the border, the CCS was doing what it could to fill the gap left by the absence of National Parks, but again there was no immediate practical result. Its 1974 report, *A Park System For Scotland*, called for co-ordinated action by all interested parties. Four types of park were proposed:

- *Urban Parks*: the town park already established but to be given an active role in recreation.

- *Country Parks* (p. 111): the mechanism for which was established under the Countryside (Scotland) Act. Thirty-six currently cover 6,426 ha.
- *Regional Parks*: along the lines of the Clyde-Muirshiel Regional Park, a place of 12,000 ha set up by the local authority in the late 1960s within which were several centres for recreation and amenity. The mechanism for official designation of such parks followed in 1981 and there are currently four, extending to 86,125 ha.
- *Special Parks*: to all intents and purposes, the same as the National Parks in England and Wales. Favoured areas were Cairngorm, Glen Nevis and Glencoe, Loch Lomond and the Trossachs – all regions mooted by Ramsay (p. 69) for special treatment. No headway was made towards such parks, however, owing to opposition from landowners and local authorities protecting their own interests and against the ever-present fear of further depopulation and loss of work.

Partly in response to the failure of the last proposal, the CCS published *Scotland's Scenic Heritage* in April 1978. This report concerned itself with forty areas of high scenic value which did not fall within the boundaries of any park or other amenity area but which, it was felt, ought to be preserved for their landscape value. The proposals were accepted and the forty National Scenic Areas (NSAs) totalling over 1 million ha (13 per cent of Scotland) received official recognition. Within them, certain planning proposals had to be vetted by the CCS, and management agreements, with discretionary payments to landowners, could be formulated by the CCS or local authorities. But their designation, like the Areas of Great Landscape Value proposed by local authorities, has turned out to be little more than a tag. With their introduction, the National Park Direction Areas (p. 79) ceased to exist, but, in real terms, NSAs have attracted neither funding nor friends in amenity or conservation, and the only efforts made are the defensive ones against inappropriate planning; efforts which should be made in any area. Nature conservation and recreation have no special part to play in an NSA and it is extremely doubtful that designation helps to preserve the scenic value of such places. MacEwen and MacEwen (1982), in fact, believed that harm had been done: 'the national scenic area is a negative, cosmetic device that widens the split in conservation between nature and scenic beauty'. Perhaps the most positive result was in drawing attention to their scenic quality – alerting some otherwise dormant senses to their spiritual value and giving them a special place in a few more hearts and minds.

URBAN CONSERVATION AND INDUSTRY

At the other end of the scenic scale come the areas of urban and industrial blight.

> The ... most regrettable impact of man on the natural environment ... is the creation of dereliction. This is truly ecological disaster.

Dereliction is an ecosystem set back to beyond its pioneer stage with little hope of return, and to such a picture is usually added the hardware of humanly created detritus.

(Darling 1971)

It is important to remember that the towns grew up because of the industry and that the Industrial Revolution was:

essentially, a rural revolution, since it changed the face of the countryside as greatly, and much more permanently [than the Agricultural Revolution]. The countryside was not so much changed as obliterated altogether.

(Vesey-Fitzgerald 1969)

With the passing of the Industrial Age, sterile eyesores became familiar, and it was not until John Barr's 1969 book, *Derelict Britain*, that the full scale of the problem was exposed. He estimated there to be 60,000 ha of derelict land in England and Wales and a further 6,000 ha in Scotland. Even the official figure for Britain of 51,000 ha was cause for grievous concern. Barr stated that the figure was increasing by four hectares a day; anything between 800 ha and 1,600 ha a year. With the coming of ECY, something had to be done.

An early effort had been made on 5,600 ha of the Lower Swansea Valley which, in 1960, was nothing more than a barren, poisoned scar. Other blighted areas soon followed suit. The Ecological Parks Trust, subsequently known as the Trust for Urban Ecology (TRUE), was responsible for a particularly imaginative project in 1977, when it set up the William Curtis Ecological Park. Here, on a restored lorry park opposite the Tower of London, the trust established a wildlife refuge and study area. Adults and children born and bred in the city were able to experience wildlife at first hand, learn new skills and begin to realise that urban dereliction can be put right after all; that the countryside can be brought back into the cities and the local spirit revitalised.

In many ways, this was the start of the urban nature conservation movement, which has since branched out into city parks, city farms, community gardens and adventure play areas, and is serviced by urban wildlife groups, community technical aid centres, the Groundwork Trusts and Community Action in the Rural Environment (CARE). It is a completely new branch of the movement, and worthy of a book in itself. It encompasses aspects of both nature conservation and recreation and yet transcends both.

Our urban improvement schemes are based on the philosophy that many of our inner city problems stem from a kind of spiritual malnutrition caused by a distancing of people from nature and we became committed to the belief that exposure to nature has an uplifting effect on the spirit and quality of life.

This passage comes from the brochure of the Landlife organisation, which promoted itself as 'Working with People for Nature; Working with Nature for People.' It was founded in Liverpool in 1975 as the Rural Preservation Association, just another countryside protection body. It soon became aware, however, that if the objects of conservation were to be achieved, large numbers of people would need to be educated in the cause. The best way in which to do this was to involve them in a practical way; the best place to do it was in the city. Urban-fringe and inner-city projects were developed and quickly expanded into a wide-ranging programme, which put Landlife amongst the leading lights of the urban conservation movement. Eighteen major projects were completed in its first ten years, sixty derelict sites landscaped and twenty school reserves created. It became a truly urban-based concern and its objects reflected the general aims of this thriving branch of the wider movement:

- to improve, protect and conserve British wildlife through practical projects involving people;
- to tackle environmental, social and economic issues through programmes that have nature conservation benefits;
- to develop a broad platform for nature conservation by bringing in considerations of urban renewal and employment creation which result in a wider class base and less reliance on temporary job creation schemes.

In addition to such earnest activity to mend old scars, there is another, slower but sure means of reclamation. The ability of the natural world to rehabilitate itself will stand it in good stead long after the last of us has died out. In 1970, owing to the closure of almost 9,000 km of railway, there were 8,000 ha of land which ought effectively to have been derelict. Instead, they became havens for wildlife, at least until sold off for other uses, acting as arteries and lifelines for creatures and plants returning to the hearts of our towns and cities. Most effects of industrialisation will eventually be turned to advantage by our wildlife if it is left undisturbed. Some of them assume amenity value, too, as human values change with time. Thus, when British Coal sought to remove Britain's largest slag heap at Cutacre Clough, Lancashire, in 1987, it faced uproar from objectors who valued the open access afforded to this famous relic of the industrial scene. They denounced the likely noise and dust from the blasting and extra traffic and argued that the site was as attractive as any of the farmland around Greater Manchester. The irony is compounded by the nature conservationists who are now calling for action to save some sites of dereliction for the animals and plants that have already occupied them. It is another question to which the countryside movement as a whole must apply itself.

Perhaps the most familiar industrial legacy today is that of the worked-out gravel pits which, left to flood, hold great potential for both wildlife and recreation. This industry, like many others, has found it worthwhile to make an effort towards restoration in order to ease the passage of later planning proposals. Throughout the 1970s, it was noticeable that applications were increasingly granted only with a condition of subsequent restoration of the landscape.

Most environmental concern shown by individual companies, however, has until now been based upon their wish to promote a favourable image. Increasingly, through the 1970s and 1980s, it became desirable for big business to be associated in some way with the conservation cause. Shell UK was one of the first to board the environmental bandwagon and it made no secret that 'the primary objective of all this was strictly business'. Shell produced posters, calendars and nature guides. In 1971, it sponsored twenty nature trails. The experiment was judged sufficiently successful to encourage repeats in subsequent years – a minimal expenditure towards a much improved image. During ECY, a competition, run in conjunction with the NC, evolved into the Shell Better Britain Campaign which became a leading initiative in funding practical conservation work (p. 187).

Various representatives from industry were involved in the Countryside In 1970 conferences. Early in the decade, major oil companies began to make voluntary compensation payments for oil pollution. They started to liaise with conservation bodies on suitable issues, and a lot of multinational giants came to employ conservation advisors. By the end of the 1970s, in the industrial arena, conservationist issues were being taken almost seriously.

GETTING PLACES

In fact, by the end of the decade, the conservationist cause was making its presence known, if not felt, in all areas of British life. To some, it was the path to a new Utopia; to others, the continuing ravings of the Doomsday prophets. To some, it was a code of living to be followed religiously; to others, an irrelevance to be ignored. But those with vested interests in the 'old order' began to feel uncomfortable with the groundswell of support for the environmental cause. Whereas, during the 1960s, they could pooh-pooh much of the Doomsday philosophy as overstated, the 1970s saw an about-face as they moved onto the defensive. The conservation movement, consisting of the 'new conservationists' as well as the nature conservationists, sounded convincing in many of its arguments, with proven facts, figures and events to support them.

The thought that there must be more sensible ways of running our affairs than twentieth century politicians and economists have so far prescribed can no longer be prevented from spreading . . .. That is particularly the case in the United Kingdom, where the most recent failures to solve national problems are superimposed on decades of economic decline which can no longer be represented as anything but a failure to perceive and adjust to the realities of modern national management.

(Johnson 1983)

Quite simply, as Dawson pointed out in 1981, 'What appears to have happened over the past decade is that the ecological view has in effect overtaken the technological'.

Here, indeed, was a change of attitudes: a change of such magnitude in one decade that the 1970s must be viewed as the most successful for the conservationists, both in the number of converts won and in the whole-sale spreading of the word. In 1971, Bonham-Carter had written: 'Professionally, it is "square" to talk of "village life". People pity or patro-nise you if you do so, or simply do not take you seriously'. Ten years later, as all the social trends confirmed, village life had come to epitomise human well-being. Those who were able to do so, sought solace there from the evils of the towns; sought comfort in the countryside's embrace. In line with these changing social values, concern increased for the future of the rural scene.

There were some who felt that concern for the environment was a luxury that we could only afford as long as we continued to grow. But recent opinion surveys show that the environment ranks third after unem-ployment and inflation as an issue that worries the public.

(Ellington and Burke 1981)

On the ground, the altered attitudes were reflected in a variety of ways. The Ministry of Defence, guardian of 272,000 ha of Britain, began to manage much of its land for wildlife in 1973, when previous lack of management was identified as a major cause of habitat deterioration. This policy came to maturity early in 1988 when the Ministry and the NCC signed a Declaration of Intent on nature conservation. The Ministry under-took to manage SSSIs to maintain and, where possible, enhance their wildlife interest, to incorporate nature conservation into the management of other areas of interest and to take sympathetic account of wildlife throughout its estate. The NCC, for its part, would advise on management plans and assist the work of ministry conservation groups. The Greenham Common cruise missile silos were turfed with natural Berkshire Heath mixture. Given the virtual absence of this habitat in the region, we must suppose that it would have failed as a camouflage ploy and that it must have been provided for the benefit of the wildlife!

In the National Parks, the divergent aims of protecting natural beauty and providing suitable recreation could not continue on an equal footing by 1976 and, when pushed on the point, the government elected to give precedence to natural beauty.

Was it possible that the politicians were 'greening'? On the international political scene, the first faltering steps were being taken. Was the seed of conservationist attitudes maybe germinating in official circles? Industry had identified a future for the cause. The minority of British people supporting it multiplied by several times and became a much increased minority. This, in turn, afforded the organised movement a much greater potential for future action. The conservationists seemed to be getting places at last. Two fundamental changes in their tactics during the 1970s account for this small success.

Firstly, they went public. This involved more than campaigns and conservation years. It meant relating the message to the everyday life of all the people. It meant making conservation matter both in the mundane routines and in those special places close to the heart. It meant using simple language. Fiascos such as the Cow Green defeat (p. 113) had shown 'the worthlessness of the specialist and exclusive approach. The conservationists withdrew to lick their, in some ways, self-inflicted wounds'. Dawson (1980) went on to show that they returned to the fray in the 1970s with 'popular' arguments pertinent to everyday existence. This new tactic paid off on occasions: for instance, against mining proposals in Snowdonia when, after a three-year battle, the mining companies withdrew. 'Environmentalism had gained a foothold' (*ibid.*).

The second change of tactics was to meet the 'establishment' bodies on their own ground, to exercise similar strategy through their accepted channels. Such moderate behaviour had been almost taboo in the previous decade, especially for the 'new conservationists'. But:

> the movement has become increasingly sophisticated over these last ten years. Instead of fighting battles on the ground, as the Ramblers did on Kinder Scout in the 1930s, they now fight in the courts, in Parliament and at public inquiries.
>
> (Dawson 1980)

The 1970s saw the nature conservationists unite with the new conservationists under a common umbrella of principles, not without some surprise in both camps. The word was spread on a flow of sound argument and proven cases. The movement matured to accept its responsibilities towards both the 'establishment' and the British public who lived under it. Dawson saw it as: 'the coming together of a movement, with the development of pressure group politics and the upsurge of environmentalism in the public mind'.

8   1980s: Building bridges and bringing down barriers

THE WILDLIFE AND COUNTRYSIDE ACT 1981

In opening our review of nature conservation in the 1980s, we could do worse than to consider the Wildlife and Countryside Act of 1981. Here, at the start of the decade, was an Act heralded as a comprehensive and effective measure to safeguard Britain's natural heritage. Here, indeed, was a concrete manifestation of how far the conservationists had instilled their message in the national conscience; how far in practice, in cruel reality, Britain was willing to go in the interests of its flora and fauna, its landscape and environment.

In Northern Ireland, the provisions of the Nature Conservation and Amenity Lands (NI) Order 1985 and the Wildlife (NI) Order 1985 have the same effect as this Act. There, Areas of Special Scientific Interest correspond to SSSIs on the mainland and, like NNRs, are designated by DoE(NI) in the absence of NCC in the province.

While the Act of 1975 had been a narrow-sighted piece of protective law, the Wildlife and Countryside Act attempted to tackle the wider requirements of conservation. It concerned itself not only with individual species, but also, and primarily, with habitats and their management. It also considered countryside access. When, on 20 June 1979, the announcement was made to promote the Bill, it was not a reflection on any change of heart amongst the politicians but an effort to bring British conservation law up to European standards. For, in spite of claims that the UK was leading the way in conservation, its legal protection of the environment sadly trailed far behind that of other EEC countries. It was sometimes slow to ratify conventions and unwilling to endorse directives. An Advisory Committee was awaited under the terms of the Directive on the Conservation of Wild Birds, and a Scientific Authority had still to be announced under the Endangered Species (Import and Export) Act. The 1981 Act would appoint the NCC to these duties.

This was the first conservation law to be passed since the voluntary movement became politically aware. It was a new experience for the lawmakers to be faced by conservationists whose demands were put in terms which they could not ignore. Consequently, the initial drafting of the Bill alone took seventeen months to complete. The change in tactics caused embarrassment to the NCC, cornered as it was between the harsh talking of the voluntary bodies and its own status of official government agency. It found itself increasingly embarrassed by a number of problems on which it found itself unable to stand firm, most notably the protection of the SSSIs. The voluntary movement soon became thoroughly exasperated by the apparent weakness of the NCC. Ironically, this sealed the ties between the various voluntary threads of the movement and united them as never before.

The bond was sealed in 1979 with the establishment of Wildlife Link (Wildlife and Countryside Link since 1993). The Council for Nature had folded up and CoEnCo, identifying the need for a body to co-ordinate the activities of the voluntary organisations, formed the Wildlife Link committee under Lord Melchett. Partaking societies included the RSPB, the RSNC, the WWF(UK), as well as FoE and Greenpeace. None of these organisations had the time, the expertise or the funds to wage effective battles in political circles and many were precluded from doing so by their charitable status. Wildlife Link stepped into the void and made the presence of the voluntary conservation movement, which could then boast a membership of about three million, very plainly felt by the politicians. With direct access to senior civil servants, regular meetings with ministers, and a hand in many reports to government departments, Wildlife Link soon became a compelling voice in political circles.

It was through Wildlife Link, and not the NCC, that the movement made its demands for the forthcoming Act, and this greatly influenced its final shape. It also succeeded, eventually, in shaming the NCC to stand up and be counted. Screaming pitch was hit when the Bill was introduced to Parliament on 25 November 1980. Such a shambles was it that the NCC, albeit belatedly, was moved to protest and, during the next eleven months, no fewer than 2,300 amendments were proposed. Vehement opposition from vested interests succeeded in keeping all but a handful from reaching the statute books, and the Act which had promised to presage a new political commitment to the environment was, in many eyes, born 'a wretched and dishevelled piece of legislation' (Rose and Secrett, quoted by Adams 1986).

Part I was traditional in its protection of species. It replaced the Protection of Birds Acts 1954 and 1967 and the Conservation of Wild Creatures and Wild Plants Act of just six years earlier. It maintained and marginally extended the protection offered by the earlier Acts. In particular, fifteen species of bat attracted comprehensive protection throughout the country, except in the living areas (only) of dwelling houses. It prohibited the picking, removal and offering for sale of sixty-two species of plant (fifty-five under the Northern Ireland Order) and allowed the NCC to collect the seeds of threatened species for storage in seed banks. In line with the EEC Directive on the Conservation of Wild Birds, it instituted SPAs (p. 142). It amended the Endangered Species (Import and Export) Act and legislated against the escape of certain non-native plants and animals. The NCC was required to review the list of protected species every five years, and at any other time it felt fit, and to advise the government of any desirable changes. Given its traditional flavour, Part I was uncontroversial.

The same was true of Part III, which dealt with access to the countryside. Local authorities were required to prepare updated definitive maps of public footpaths in their areas and they were empowered to arrange access or management agreements with landowners.

But Part II, relating to habitat conservation, heralded much that was new to countryside interests and hung it on a bone of contention upon which the nature conservationists have gnawed bitterly in their frustration. The meat on the bone allowed the NCC to declare NNRs on suitable land held by other approved bodies and to award grants and loans to suitable conservation schemes. Importantly, it made provision for official MNRs between high water mark and the three-mile (five-kilometre) territorial limit. It also gave some protection to the floristically rich limestone pavements. But this was lean meat on a bone which has served to widen dramatically the divide between conservationists on the one hand and farmers and foresters on the other. For the provisions relating to habitat protection, specifically the SSSIs, were an unworkable mess.

Habitat conservation had become the new standard for the movement and it was an urgent priority to get it right.

> today habitat destruction is proceeding at such a pace that the creation of reserves, however effective twenty or thirty years ago, is now plainly inadequate. Most national parks and nature reserves are much too small .... Conservationists simply must find some way of conserving a substantial proportion of the world's remaining unspoiled ecosystems without waiting to see how these areas are going to be named, administered or used.
>
> (Fitter and Scott 1978)

> Site protection ... is neither an adequate nor a very appropriate approach to the conservation of the wider countryside.
>
> (Mabey 1980)

> there is a danger of assuming that the objects of conservation can be fulfilled simply by creating reserves ... we also need a much broader approach as well to make certain that those who own and occupy and use land regard conservation as part of their responsibility. Conservation cannot be relegated to reserves alone and requires a conscious and continual effort from all those who care for the land.
>
> (HRH Prince of Wales, in Hywel-Davies and Thom 1984)

On the face of it, Britain's countryside was sinking under the weight of protection. The book from which the last quote was taken listed 2,000 reserves covering 222,000 ha. Over 200 of them were NNRs. Then there were 4,000 SSSIs extending up to 1.4 million ha, with another 1,250 sites in the pipeline to cover a total of 8 per cent of Britain by 1986. National Parks covered 9 per cent of England and Wales. AONBs accounted for 10 per cent of England and Wales and 18 per cent of Northern Ireland. NSAs applied to 12.7 per cent of Scotland. By 1989, Green Belts covered 12 per cent of England (1.5 million ha) and, with Areas of Great Landscape Value (AGLVs), 12 per cent of Scotland.

But many of those areas overlapped; many were concentrated in small geographical pockets. Designation rarely conferred protection. Even in the NNRs, the NCC was able to spend no more than £15–20 per hectare each year on management. Elsewhere, nature conservation and planning control were largely a matter of voluntary contribution; of wheedling and willing, harrassment and hope; half-measures and mixed fortune. All the cosmetic designations combined could cover less than a quarter of the area under farm and forest. Wider habitat conservation was vital.

Of Britain's 22.7 million ha, 77 per cent was devoted to agriculture and 10 per cent to forestry, compared with 57 per cent and 25 per cent, respectively, for the EEC as a whole. Arable area in England increased only from 4.5 million ha in 1955 to 5.2 million in 1984 but productivity rocketed due to intensification. Some 1.3 million ha remained under rough grassland while 3.2 million were 'improved' pasture. All the manifestations of modern agribusiness – mechanisation, pesticide use, reseeding of pasture, clearance of marginal areas, removal of woodland and hedgerow, infilling of ponds and ditches – all were having dire effects over these vast tracts of Britain.

Of the 2.5 million ha devoted to forestry, just over 2 million ha were productive forest, the figure projected in 1943 for the end of the century! A further 170,000 ha were classed as unproductive woodland. Private interests controlled over half of this productive timber. Further land, which in the case of the FC amounted to 276,000 ha, was new or proposed plantation or devoted to peripheral use. In 1983, the FC planted 9,000 ha, and private foresters 12,500 ha; an average year's planting for the 1980s. The FC was looking to double its planted area to a total 1.8 million ha by the year 2025.

Of the productive woodland, 70 per cent was coniferous, (94.5 per cent in the case of the FC) and fewer than one in ten trees being planted was broadleaved.

Like farming, forestry has far-reaching consequences for the environment. It alters the flora and fauna, often quite drastically replacing one set of organisms with another. It changes the properties of the soil and acidifies stream water. A forest can reduce water run-off by 15 per cent over that of grassland. In terms of amenity, it has drastic, eye-catching effects upon the landscape and can often restrict access. It conflicts with other land-uses, most particularly hill farming and upland sporting interests.

Farming and forestry have always moulded the rural environment. What the conservationists seek now to preserve is a result of their past operations. But, by the mid-twentieth century, the pace and scale of these operations was sweeping all before it, with extravagant, sometimes unexpected results which were then unquantified and to which no limit could be seen. It was the farmers and the foresters who needed to be convinced of the utter necessity to conserve habitats. If they would not do so, all the cosmetic designations would fail.

The NCC had always been duty-bound to notify the planning authorities of any SSSI in their area. SSSI status precluded any consideration of planning development without first consultation with the NCC. It was known that such status afforded no real protection but at least the pretence was shown. Efforts to tighten up the protection of SSSIs had failed in Parliament in 1964 and again in 1968. The calls of the 1965 Countryside In 1970 conference for planning controls on agriculture in SSSIs went unheeded. Protection remained a voluntary option of the landowners. The NCC kept the list of sites under constant review and, by 1975, 2,422 sites covering 944,715 ha had been notified in England and Wales and 787 (550,345 ha) in Scotland. As a manifestation of the lack of protection, though, 113 former sites had lost SSSI status by that time and a further eighty-seven had been reduced in area. When the Wildlife and Countryside Act was introduced, 3,800 SSSIs covered 5.3 per cent of the UK, but about 100 sites were being damaged every year, some of them irretrievably. The conservationists had high hopes that some real protection would be forthcoming. But long before the Act became effective, they realised that their hopes were founded on sand; the landowners were too powerful for the politicians, and voluntary action or inaction would remain the order of the day.

The NCC now found itself faced with an immense task. The Act required that all those sites notified to local authorities prior to 30 November 1981 were to be renotified to the owners and occupiers of the land. 'In the interim, owners and occupiers who are aware of the existence of an SSSI on their land are asked to follow voluntarily the procedure ... in relation to any operations which might be damaging to the scientific interest of the site' (DoE 1982). Alternatively they might choose to damage it before the notification process was complete. Protection, in short, was non-existent. Furthermore, the NCC was still required to notify new sites, and a minimum three-month consultation period then followed to allow for landowners' representations to be considered. It was an extremely lengthy process.

Once an SSSI had been notified, the landowner was prevented from carrying out any of the operations listed until he had given written notice of his intentions. The NCC then had three months in which to agree to the operation or to arrange a management agreement with the landowner. If they failed to do so within that time, the operation could go ahead.

The number and variety of Potentially Damaging Operations (PDOs) shows just how vulnerable SSSIs are to modern techniques of land management. They include:

- cultivation (ploughing, reseeding)
- grazing
- mowing
- fertilising
- pesticide use
- dumping
- burning
- release of any species (wild/feral/domestic)
- woodland management
- drainage/alteration of water courses
- infilling of ditches, ponds, etc.
- freshwater fishery production / coastal fish management
- reclamation of land from sea/marsh, etc.
- bait digging (England and Wales only)
- erection of sea defences
- mineral extraction
- road/pipeline construction/removal
- storage
- building work
- modification of natural/man made features
- geological collection of specimens
- use of vehicles
- recreational activities
- game/waterfowl management

Where grant-aid for an operation in farming or forestry was refused on conservation grounds, the NCC was required to offer a management agreement. Thus, the NCC would be compensating farmers and foresters for loss of potential grant monies. Alternatively, the landowner could decline the offer of an agreement and go ahead with the work at his own expense.

The Act did allow the conservationists one final hope in extreme cases. The NCC can ask the Secretary of State to make a Nature Conservation Order (NCO) on a threatened SSSI, but only if 'for the purpose of securing the survival of particular kinds of plant or animal, of complying with an international obligation or of conserving species or features of national importance'. The landowner then has twenty-eight days to lodge an objection, but the grand title of NCO merely extends the consultation period from three to twelve months. To compound the frustration of the conservationists, time has shown that not all SSSIs are thought to be eligible for NCOs; the opinion of the ministers has been that only some sites are of sufficient value to warrant this extra 'safeguard'. If a management agreement has not been made successfully by the expiry date of the NCO, the landowner is free to go on with his planned operation. The NCC's only recourse then is to its power of compulsory purchase under the 1949 Act. It is a power which the council was understandably loath to use, and only in 1990 did the NCC make application for its first purchase order after more than four years of negotiation had failed to save part of the Westhay Moor SSSI in Somerset from nothing more than invasion by scrub. The landowner had declined to control the invasion to save many rare and interesting plants.

Another indication of the Act's impotence was the punishment that could be brought upon the person carrying out an operation without permission: a fine of up to £500. The Act appeared only to take at all seriously the contravention of an NCO which could attract a fine of up to £1000 plus a further £100 per day until the land was restored to its former condition. The NCC was empowered to do this work and to recover its expenses from the person responsible.

To the injury caused by this complete lack of protection was added the insult to conservationists of having to pay landowners to act on behalf of wildlife. Their fear was that landowners might start to plan operations which they would otherwise not have considered. Who, after all, would protect the status quo of an area for nothing when they could receive payment for doing so? Landowners, now alerted to a new potential source of income, were bound to seek advantage from it. And compensation based on theoretical losses sounded very fanciful indeed.

But the movement might have known that it would have to start paying its way before long. Just as it had become politically aware, so too did it need to learn the facts about economics. In the 1980s, voluntary agreements could only succeed where compensation might be paid. Any new demands for statutory moves to protect the environment would likewise bear a price which the conservationists would have to pay. But, viewed positively, the monetary angle was not all gloom. After all, the Wildlife and Countryside Act made central funds available for conservation work. The British people, through their government, were subsidising conservation for the first time in its history. The real problem was not the fact that conservation had to pay its way, but that the official government body was so severely limited in what it could do by lack of funds. By far the lion's share of any compensation payments would go to the farming and forestry communities. Yet, while MAFF Improvement Grants in 1978 ran to £540 million, the total budget from which the NCC had to fund all its work was £7 million!

In that light, nature conservation was not an attractive proposition. But now that government funding for conservation had been accepted in principle, the way was open for far greater things. If the movement could increase its funding, it could also, most certainly, increase its attractiveness to landowners, and those voluntary agreements would follow so much more easily. 'Agricultural subsidies when crudely and unimaginatively applied have done more to damage the flora and fauna of Europe than any other single factor' (Moore 1987). Here was a chance to change this historical fact. The movement would have to encourage its government to fund it realistically; then, with money in the bank, it could sell nature conservation as a valid option for the landowner. A small start was made in 1983, when the government, having received particularly bad press over the destruction of the Somerset Levels, began to finance all management agreements on SSSIs from central funds.

It was the injury caused by the lack of real protection for habitats that hurt the conservationists most sorely. Some occupiers, hearing of NCC intentions to declare SSSIs on their land, purposely damaged the sites before notification could be made. So time-consuming was the process that it was not fully completed until 1991, but to the credit of NCC staff, this represented a mammoth amount of work.

The tragic loss of valuable habitat over the years is an indictment of the greed of the landowners and the unconcern of the politicians but, as the clamour intensified, so too did the scale of the damage. During the year April 1983 to March 1984, 156 notified or intended SSSIs were damaged, sixty-seven of them seriously. The worst offences against them are listed in Table 8.1. By the end of the year, 46 SSSIs had been denotified. During the following year, ending March 1985, 255 sites were damaged, ninety-four of them seriously and 137 of them long-term. By then, the NCC was expecting to have to denotify a total of 266 sites in England alone – 10 per cent of the total number then in existence. Clearly the principle of voluntary conservation was failing. But was it all due to the greed of the landowners and the unconcern of the politicians? Did the conservationists perhaps persist in remaining too aloof from the mainstream? Was their message still too remote from day-to-day realities? Why did the average landowner not feel favoured to possess an SSSI; why no element of excitement in the natural richness of the ground? Money talks, and economics rules in modern Britain but, even in farming and forestry, they need not outweigh a well-reasoned argument based on other principles. The Act required MAFF to advise landowners on matters relating to conservation and amenity. Until answered, it must remain a burning question among conservationists why so little was willingly done for their cause by those outside the movement and why they were themselves so little involved in drawing up official guidelines after the Act was passed – guidelines on SSSIs, compensation and management agreements. Why were they left out in the cold? Were they viewed with distrust, or not viewed at all?

The stark truth exposed by the Wildlife and Countryside Act was that the British nature conservation movement had still not made the grade. It had instilled its message into the national conscience, but only the

Table 8.1 Damage to SSSIs 1983–4

Cause of damage	No. of sites
Woodland management	31
Drainage/alteration of water courses	28
Cultivation	18
Dumping	15
Fertilising and pesticides	14
Grazing	9
Burning	8
Recreation	5
Infilling of ditches/ponds etc	4
Mineral extraction	4
Road/pipeline construction	4

conservationists themselves were paying it more than lip-service. The politicians viewed it as a fringe interest; 'there are no votes in conservation' (Peter Walker MP). The landowners would not afford it; the economists could not make it count. Having come of age in the previous decade, conservation was still too private and remote; its exponents too narrow-minded and self-centred to carry conviction.

Yet some of their claims could not be gainsaid, and the public outcry over loss of wildlife habitat through farming increased in intensity. It was something to which the public could relate. The argument that habitat change affected wildlife populations was a simple one to follow, and many people had first-hand sight, at least, of its effects. The politicians began to listen and the farmers came to realise that their thirty-five years in clover were coming to an end. When the all-party House of Commons Environment Committee reported on Part II of the Act in January 1985, nine of its eighteen recommendations related to agriculture. It suggested a government review of the overall use of 'the rural estate'. It proposed that conservation be given comparable status with food production and that MAFF should bear this in mind when awarding grants.

Hot on the heels of this inquiry, Dr David Clark MP promoted a Private Member's Bill calling, in particular, for stronger laws to protect SSSIs and for a duty towards conservation to be imposed on farming and forestry. But the Bill met with vehement opposition from vested interests and, when it became law on 26 August 1985, the Amendment to the Wildlife and Countryside Act was as great a disappointment to the conservationists as the Act itself had been. The time allowed to the NCC to thrash out management agreements on threatened SSSIs was increased from three to four months. SSSIs were protected from damaging operations from the moment the owner received notice of NCC intention to (re-)notify the site. The proposals to impose a conservation duty upon forestry had become nothing more than a requirement that the FC seek a balance

Plate 8.1   Farmers were not the rogues that many took them to be. They had a job to do and MAFF was making a better case than were the conservationists. What was needed were new initiatives with a positive message: farm trail at Stoneleigh, Warwickshire. Courtesy of Countryside Commission

between timber production and wildlife conservation – and upon the farmers, nothing. Wildlife and habitat conservation remained irrelevant to the food factory of the British countryside. Such was the power of the farmers' political elbow. It was a fitting amendment to an impotent Act.

FIGHTING TO BEFRIEND THE FARMERS

Nevertheless, the farming fraternity was becoming concerned at what it saw as a new national sport of 'farmer-bashing'. The year 1980 saw the publication of Marion Shoard's *The Theft of The Countryside*, as effective an item of propaganda as Rachel Carson's *Silent Spring* had been in its day. Richard Body joined the fray with *Agriculture: The Triumph and The Shame* (1982) and *Farming In The Clouds* (1984) – and to make matters worse for the farmers, he was a Conservative MP. In tactical response, the two power-machines of agriculture, the National Farmers' Union and the Country Landowners' Association published reports in 1984 which tended to suggest that there may be a place for conservation in the activities of their members. In the same year, MAFF set up an environment co-ordination unit and began to review the activities to which it might grant financial aid.

In truth, the ministry's about-turn was entirely due to public pressure channelled through voluntary conservation groups, such as the Royal Society for the Protection of Birds and the Council for the Protection of Rural England. They had proved far more effective lobbyists than the civil service.

(Rose, in Goldsmith and Hildyard 1986)

Again, the conservationists are being somewhat inward-looking here; their self-congratulation is blinding them to the truth. It was hardly an about-turn by the ministry; more a subtle foot-shuffling due to changing circumstances. At the same time, for instance, in order to remain free from EEC conservation-based regulations, it was questioning whether or not the EEC could legally include environmental considerations in agricultural requirements. On the continent, the ministry lost the argument, but we have already seen how, in British law, it was able to fend off conservation duties. Any move towards conciliation was due as much to factors within the farming industry as to any public pressure. Surpluses of various products – the lakes and mountains of the modern farming scene – are perhaps the most obvious. The resultant quotas, falling grant money, reductions in guaranteed prices were all conservationist by default, not by design. The 'set-aside' scheme to pay farmers to take land out of production might affect up to 250,000 ha, 5 per cent of Britain's arable land. Its conversion to grassland, woodland or other uses might have far-reaching implications for conservation, but only if the farmers could be encouraged to consider wildlife and amenity. The land was their livelihood and they would still expect some return from it. Would the conservationists act correctly and quickly enough to convince them this time (p. 242)?

The alternative to 'set-aside' would have been to seek ways of reducing the intensity of farming methods. On the face of it, such extensification would seem extremely beneficial to wildlife. But we would be wrong to presume that it would mean a return to small hedge-lined fields clothed with wild flowers and rough corners, copses and ponds alive with wildlife. Those farming days are gone. There is little point in dreaming of the benefits that may have come from extensification. That course would have been prohibitively costly to farmers; lip-service that expensive was not about to be given to a cause that remained ethereal to most landowners.

The farmers still held all the cards. While the official representatives of both sides maintained a cautious dialogue in search of common ground and while many farmers, of their own volition, were making genuine efforts on behalf of wildlife, the conflict in other quarters became deep, divisive and damning. On 22 February 1983, farmers at West Sedgemoor in Somerset publicly burnt effigies of NCC officials in protest at the SSSI designation of the moor. In the following year, similar effigies were hung from gibbets in the Orkney Islands. In June 1984, a Norfolk farmer was blasphemous in another way. His land in the marshes of the Norfolk

Broads was a haven for birds, dragonflies and plants. But for him, its potential was in cereal production, after he had drained it with the help of lucrative grants. Aware of the concern of the conservationists – aware also that they were completely and utterly powerless to stop him – he ploughed a large 'V' across one of the fields as a first step in the operation. He left the conservationists to interpret the sign as they saw fit.

They interpreted such actions as savage and sacrilegious. Agriculture was still victorious in the British countryside, and the gap between farmer and conservationist was probably wider than it had ever been, even though most hostility was rather more subtly expressed. Bridges had to be built if Adams's (1986) image of future British land use were to be avoided:

> The North and West would become a land where agriculture exists primarily to maintain wildlife habitats and landscape beauty, and farm tourism and crafts dominate the economy. In the lowlands, however, the business of intensive production will go on, even if on a restricted area. The incompatability between conservation and intensive agriculture will then develop a stronger regional dimension ... it is exactly what policies of conservation site protection are themselves tending to produce.

Should this picture become a reality, then we shall surely know that the conservationists failed to reconcile with the farmers.

Lip-service from agriculture continued in the form of Environmentally Sensitive Areas (ESAs). This was a political response to increasing public pressure and to a damning 1984 House of Lords report, *Agriculture and the Environment*. An ESA is a 'particular area of recognised importance from an ecological and landscape point of view' in which agricultural practices should be 'compatible with the requirements of conserving the natural habitat and ensuring an adequate income for farmers'. To this end, MAFF awards grants to landowners who attempt to preserve characteristic landscapes by continuing or returning to traditional practices. Farmers enter the scheme on a purely voluntary basis.

An EC regulation introduced the ESA concept in March 1985, and the machinery was operating in Britain by the end of that year. At the same time, a clause was added to the Agriculture Act, placing a duty on farmers to pay due regard to conservation. It was this very duty which they had fought so hard, and successfully, to keep out of the amendment to the Wildlife and Countryside Act just a few months before.

The NCC and Countryside Commissions identified forty-six potential ESAs in England and Wales, including all the National Parks, and a further seventeen in Scotland. Pruning their expectations, they presented to MAFF a priority list of thirteen areas south of the border and five in Scotland. These were declared before the end of 1987, as was one in Northern Ireland. They totalled over 800,000 ha. A total of £8 million a year was made available for their funding and 2,500 farmers applied to enter the scheme during the first eighteen months. Some regions acted more positively than others. In England, over 75 per cent of eligible land was entered but a poor response from other areas suggested that modernisation was continuing unabated in such places. Since farming practices in these regions remained relatively traditional, it was here that wildlife still had most to lose.

As with any new initiative, there were both the doubters and the enthusiasts.

> In reality, ESAs will be merely another form of farm grant and there will be no planning powers to stop destruction of landscape features or wildlife habitats. This palliative will neither stem the haemorrhage of Britain's wildlife heritage, nor create an economic and ecologically acceptable form of farming which could promote harmony not only between farmers and the landscape and its wildlife, but also between farmers and conservation groups, as the government claims it wants to do.
>
> (Rose, in Goldsmith and Hildyard 1986)

David Conder of the CPRE felt that the ESA:

> proved the high point in co-operation between the two green forces. Conservationists welcomed it for two reasons: firstly MAFF was taking direct responsibility for funding conservation-sensitive farming – costs that had previously been borne by the relatively tiny conservation authorities; secondly, for the first time, income support for farmers can be partly divorced from increased food production and will be made conditional on environmental good practice.

The fact was that the ESA scheme provided the machinery for agriculture to help conserve wildlife and habitat. The factors that would affect its overall success or failure – the level of funding, the number of ESAs, the enthusiasm of the farmers – now depended upon public support for what the scheme endorsed. Perhaps the ball was back in the conservationists' court?

Meanwhile, both the amenity and nature conservation wings of the movement were increasing their calls for planning controls on farming and forestry. This was one of the banners carried so ably by Shoard (1980). It dismayed conservationists that the town and *country* planning machinery was so powerless against the forces that shaped the country scene. The problem arose because, ever since their governing Acts of the late 1940s, planning, agriculture and conservation had been allowed to develop in their separate ways. Agriculture in 1947 posed no threat to the countryside; it was the countryside. Agriculture was not responsible for the eyesores such as ribbon development which the planning machinery was set up to tackle. The farmers were the friends, the benefactors of the conservationists.

Plate 8.2 In the ESAs, agricultural practice should both conserve nature and ensure the farmers' incomes: lip-service continues at Swaledale in the Pennine Dales ESA and Yorkshire Dales National Park. Photograph by W. Meadows. Courtesy of Countryside Commission

But times changed. Farming, like forestry, altered the scale and the intensity of its operations. The effects of these operations became increasingly long-term and often altogether final. The natural step now, as far as the conservationists were concerned, was to bring farming and forestry within the provisions of the planning machinery, not to obstruct their operations, but to allow consideration of them alongside other rural interests before they were imposed. Surely, it was felt, some sort of arbitration ought to be available before a hedgerow was removed, for example. The alteration of a drainage course or the bulldozing of a copse was, after all, more final than the erection of an inappropriate building which could always be removed. Even within the confines of building development, it seemed somehow sinister that a landowner could put up a steel barn without planning permission and then sell off traditional buildings at immense profit as potential dwellings.

Out on the commons, the battle of preservation was being fought against the failings of the 1965 Commons Registration Act (p. 110). As the Act had failed to specify whose duty it was to register common lands, many had not been notified before the 1971 deadline and had effectively become the private estates of their owners. After the simple step of deregistering the land, the owners were free to make any number of changes that had previously been impractical due to commoners' rights. Some commons were sold for development, as at Copthorne on the Surrey/Sussex border. Others became grouse moors, as at Keldside in the Yorkshire Dales National Park, or even golf courses, as at Walton Heath in Surrey. The Shotton Paper Company afforested the 82 ha of Cefn Coch Common in Gwynedd. Farming commandeered a great many of the forfeit commons. And nothing could be done to stop it, thanks to loopholes in the law that should have secured the future of the commons. Only where landowners jumped the gun of deregistration was there recourse to action and, even then, local authorities seemed reluctant to act. Not until 1986 was the first positive move witnessed when Dyfed County Council sought a High Court Order to prevent fencing by landowners on Plumstone Mountain near Haverfordwest.

In order to overcome some of the problems and to seek an agreement between the interested bodies on a basis for legal protection, the CC – which had been given its independence from the DoE by the 1981 Act – set up the Common Land Forum in 1984. This involved the farming associations, local authority representatives and the amenity interests led by the Open Spaces Society. Its report of June 1986 recommended a clause to prevent deregistration, continuing public access with a duty upon county councils to remove unlawful fencing, and the setting up of management associations to involve all interested parties. Management would aim to promote the conservation and enhancement of natural beauty, encourage public access and maintain high standards of livestock husbandry. Initial government response appeared sympathetic, with the 1987 election manifesto containing a promise of action. Three years later, on the final day before its summer recess, Parliament announced a U-turn and deregistration continued into the new decade. The courts have prevented some attempts to deregister, but it is an expensive and unsatisfactory situation for all concerned, and one that needs addressing positively.

FORESTRY, FISHING AND FORESHORE

The forestry industry had never sought such open conflict with the conservationists as had some sections of the farming community. Of all the unpopular patterns of planting, there was never a 'V' across an open hillside or moor. There was no very logical explanation for these closer ties, unless we credit the FC with more discretion and subtle tact than MAFF. For most forestry was carried out by the commission, or by private foresters who accepted its principles and followed its guidelines. Yet forestry had as much potential as farming to alter landscape and habitat, and the predominance of conifers in its schemes did nothing to enrich the value of either. Vast areas had been affected: 11.2 million ha of open heath, moor and peatbog since 1918, for example. Nor was it all confined to upland regions: 95 per cent of woodland in Dorset was deciduous in 1811; by 1972, this had decreased to 23 per cent. Yet, apart from early hiccups over the aesthetic barrenness of some plantations, forestry had enjoyed an easy ride with the conservationists. Consequently, calls for planning controls on forestry were sometimes tempered with an alternative suggestion based on licensing new plantations. The movement, it seemed, trusted forestry to act with some consideration for its cause. But this trust was about to evaporate.

The FC, in accordance with the Forestry Act 1967, performed two duties in advising government on forestry policy (the Forestry Authority) and managing the nation's woodland estate (Forest Enterprise). Owing to the obvious clash of interests, the two were split in April 1992.

As Forestry Authority its objectives are:

- to advance knowledge and understanding of forestry and trees in the countryside;
- to develop and ensure the best use of the country's forest resources; and to promote the development of wood-using industry and its efficiency;
- to endeavour to achieve a reasonable balance between the interests of forestry and those of the environment [this was as a result of the Wildlife and Countryside (Amendment) Act 1985];
- to undertake research relevant to the needs of forestry;
- to combat forest and tree pests and diseases;
- to advise and assist with safety and training in forestry;
- to encourage good forestry practice in private woodlands through advice and schemes of financial assistance and by controls on felling.

As Forest Enterprise, the commission's objectives are:

- to develop its forests for the production of wood for industry by extending and improving the forest estate;
- to manage its estate economically and efficiently, and to account for its activities to Ministers and Parliament;
- to protect and enhance the environment;
- to provide recreational facilities;
- to stimulate and support employment and the local economy in rural areas by the development of forests, including the establishment of new plantations, and of the wood-using industry;
- to foster a harmonious relationship between forestry and agriculture.

A parliamentary statement in December 1980 outlined its immediate plans for the decade:

- a continuing expansion of forestry is in the national interest;
- there is scope for new planting to take place at broadly the rate of the previous 25 years – about 30,000 hectares per annum;

- there should be a greater proportion of planting by the private sector, but with a continuing Commission programme of planting;
- a proportion of the Commission's woodlands and land awaiting planting should be sold, in order to reduce the Commission's call on public funds.

The FC was the largest public landowner in Britain, with 1,165,000 ha or over 6 per cent of the land surface to its name in 1987. For economic reasons, there had been a lull in planting late in the 1970s, but this had passed and planting looked set to increase throughout the new decade. Forestry policies were thus of great interest to the conservationists. But until the 1985 Amendment Act, the FC had made its overtures primarily to the amenity wing of the movement. Discomforted by the bad feeling aroused by the regimental planting of early stands of trees, it was careful to see that new plantations were made less symmetrical; they were edged with different varieties and displayed some sympathy with the slopes and lie of the land. Public access for recreation was encouraged (p. 109) and the FC had generally shown concern for the visual integration of its operations into the countryside.

Now, in 1985, the nature conservation wing of the movement had caught up with it. Again the FC acted correctly and 'welcomed this new duty as a reflection, in statute, of the policy that it has been pursuing for very many years' (FC leaflet). A month ahead of the Amendment Act, the FC announced plans to encourage more deciduous planting both in its own operations and by private landowners through a Broadleaved Woodland Grant Scheme which began in October 1985. Now it appointed a Nature Conservation Consultant and rapidly revised its literature. The following extract is from *The Forestry Commission and Conservation*, published in March 1986:

> The Forestry Commission recognises that the conservation of nature is an integral part of forestry, and the general aims of its conservation policies are;
>
> 1 To conserve by appropriate management scheduled sites of special nature conservation interest within the forest estate.
> 2 To enhance the nature conservation value of the forest estate as a whole.
> 3 In all forest or nature conservation activities to pay due regard to wider environmental values.
> 4 To promote the public use of areas of conservation interest where this is compatible with other conservation objectives.

At the same time, it drew itself into close consultation with the NCC and the voluntary bodies on the furtherance of its nature conservation policy. An agreement on the management of 344 SSSIs covering over 70,000 ha was signed with the NCC in June 1986, and a number of new initiatives with the voluntary movement were started. A bonus in all this for the conservation movement was that many private foresters followed the lead of the FC. But there was a black cloud on the horizon and a storm in the offing.

The flow country of the extreme north of Scotland is a desolate, wind-swept region. Its 400,000 ha of blanket bog literally floats above the bedrock and moans to the sound of breeding divers and wading birds. It is the largest mire in the world and the most extensive piece of single habitat in Britain. It has been nominated as a World Heritage Site (p. 192), to put it on a par with the Brazilian rain forests. Whether or not it is beautiful is a decision for the individual observer, but its value as a wildlife habitat is unquestionable. The foresters, however, began to view it as a potential plantation of massive proportions. It was less controversial than good farmland – or so the industry thought. It was less protected than the National Parks and far enough from most of the country's residents not to be well-known and valued in its natural state. It also lay in an area of dire unemployment.

By the end of 1988, 60,000 ha (15 per cent) had been afforested, 35,000 ha by the FC. Up to 12,000 ha more had been earmarked for planting, with the FC at that stage seeking another 30,000 ha at least. But by then, the flow country had become a political hot potato. In 1987, the RSPB called for a complete stop to afforestation of the area. The NCC, just a few months later, more moderately proposed a moratorium on planting until an integrated agreement could be drawn up. To placate the conservationists, a new SSSI of 172,000 ha was announced in January 1988. But when all sides came together in the following month under the auspices of the Highland Regional Council, the scale of their differences was exposed. Nothing even purporting to be an agreement was forthcoming, and there remains a real danger that the conflict might escalate into an all-out war similar to that between the farmers and the conservationists. If that happens, all the good work of the past will pale into insignificance.

The conservation movement was also finding itself drawn into the affairs of the fishermen. Under the 1981 Act, the Secretary of State was empowered to designate MNRs up to three miles (4.8 km) from the shore. It had long been a concern of the movement that pollution, collecting and disturbance by divers was threatening the vulnerable and understudied communities of the shallow seas. Thus it was with some urgency that plans were pursued to set up marine reserves under the Act. But, in every area, opposition was encountered from local interests – mainly fishermen – and the Act was quickly shown to be ineffective in such a climate. For the NCC could only apply for designation if there was no opposition from local interests. Clearly, the marine environment, like the farming one, would have to rely upon voluntary agreement for its protection. Voluntary reserves were designated at St Abb's Head, Roseland, Wembury, Purbeck, Strangford Lough, and the Islands of Bardsey, Lundy and Skomer. Not until legal provisions were tightened under the 1985 Amendment was it found practical to designate the first official MNR – around Lundy in 1986. In spite of widespread consultation with all local interests, the NCC did not achieve a second marine reserve, this time around Skomer Island, until 1990. Inevitably, most progress continued to be made by voluntary agreement; for example, the MNR at Seven Sisters in Sussex, announced in June 1987.

Relatively few people are fortunate enough to enjoy the shallow sea environment at first hand, and consequently its main champions were the diving fraternity and some scientific conservationists. The coastline is closer to most hearts and it was up on the cliffs and beaches that the fight for preservation was pursued most avidly. By 1984, the NT's Enterprise Neptune (p. 114) had raised over £7 million and protected more than 40,000 ha of land along 725 km of coast. It was relaunched in 1985 and, three years later, with over £10 million to its name, purchased its '500th mile' (800 km) for just £1. This nominal figure was paid to British Coal for a stretch of beach at Horden in County Durham, used for centuries as a dumping ground for colliery waste. By 1995, another 50 km had been added.

Seeing the success of Enterprise Neptune, the government of the late 1960s had been moved to review its attitudes towards coastal development.

Plate 8.3 Official recognition of our best coastlines belatedly arrived with Heritage Coast designation: Flamborough Head Heritage Coast. Photograph by Ian Carstairs. Courtesy of Countryside Commission

On the advice of the National Parks Commission, it announced the selection of 'Heritage Coasts', to be nominated by local authorities and managed in conjunction with the commission. By 1995, there were forty-five such coasts in England and Wales, covering 1,523 km or almost a third of its coastline. North of the border, 7,520 km (74 per cent) was designated conservation zone. The Heritage Coast Forum, which had looked after the designated coasts of England and Wales, was replaced by the Coastal Heritage Forum, which was to care for all the coastline of Britain, with particular reference to the wider, undeveloped areas.

For the pressures on the coastline were as great as ever and neither those pressures nor the wildlife communities they threatened were sufficiently understood. The NCC launched its Coastwatch survey in 1987 and, on its completion five years later, its successor, English Nature (EN), replaced it with the 'Campaign for a Living Coast', in response to an expected loss of 10,000 ha of shoreline in the coming twenty years. The Marine Nature Conservation Review was to be an inventory of all 163 British estuaries, and the Seasearch survey of the Marine Conservation Society (MCS) was looking at underwater habitats. The House of Commons Environment Committee considered coastal zone protection, and planning guidance followed from the DoE. In 1992, the Coastal Research and Management Group, a forum for scientists and site managers, assumed the role of the UK branch of the European Union for Coastal Conservation. At about the same time, the EC was funding the BioMar project to map and fully describe the marine and littoral environment.

They were all urgent projects. The gaps in our knowledge were alarming and it would have been foolish to expect success in conserving something we knew so little about. EN (1993) complained that 'marine conservation is about a decade behind its land-based counterpart'. In reality, it was about five decades adrift. With 500,000 ha of marine habitat identified as needing active conservation, Britain had just two statutory and a handful of voluntary MNRs. If land management was piecemeal, treatment of the shoreline was thoroughly unco-ordinated.

INTERCEPTING INDUSTRY

Meanwhile, EC directives on the quality of bathing waters and of shellfish waters were causing embarrassment in Britain. Only a handful of beaches had reached required standards of cleanliness when the directive came into effect in 1975, and the problem of pollution, mainly by raw sewage, was so great that the politicians had been forced to sidestep the issue by claiming that just twenty-seven beaches were used traditionally for bathing. Such uproar did this generate from groups like the MCS and the Coastal Anti-Pollution League that a further 350 bathing beaches were acknowledged in the mid-1980s. But in spite of campaigns to get the coast cleaned up, very little effective work was done.

Consequently, an EC Blue Flag Award scheme for clean beaches passed only seventeen of more than two hundred considered in 1987. The remainder either lacked facilities or, more often, showed signs of pollution from sewage, seaweed, litter, oil or industrial discharge. In 1988, the government earmarked £70 million p.a. to tackle the task, but this amount was neither enough to be effective nor sufficient to convince the EC of Britain's good intentions. Official claims that 76 per cent of the 440 designated beaches were already up to standard failed to impress. An infringement procedure was started for non-compliance with the directive on bathing waters, when it became clear that the planned course of UK action was unlikely to bring its coasts within EC guidelines before the closing years of the century.

Marine pollution also became a major issue during the 1980s, particularly in the North Sea. It was an international problem in which Britain was inevitably involved as she relied heavily upon the sea's resources and yet was implicated as a major polluter. It was unsurprising that the problem of marine pollution should have its first obvious effects in the North Sea, a limited mass of water largely enclosed by industrial nations. Every year, it was expected to swallow 3.8 million tonnes of industrial waste; acids and chemicals, metals and shale. 65,000 tonnes of heavily contaminated dredge spoil and five million tonnes of treated sewage were also thrown in for bad measure. Britain was the only nation to use the North Sea as a lavatory, and sewage not previously treated went in as it was. Rivers disgorged industrial pollutants and chemicals from agricultural pesticides and fertilizers. Radioactive cooling water from nuclear establishments was discharged by pipeline. Oil seeped from rigs and ships in the normal course of events, during tank cleansing and at times of accident. Toxic chemical wastes, such as PCBs, were burnt off on board incinerator ships at sea, and acid rain (p. 186) fell on the ocean as well as the land. In the 1980s, it all added up to about 70 million tonnes, containing 50,000 different chemicals, every year.

Various conventions existed to tackle one aspect or another of the problem. Most important for the North Sea were the Oslo Convention of 1972, dealing with incineration and the dumping of sludge at sea, the MARPOL (marine pollution) Convention of 1973, and the Paris Convention of 1974 which controlled discharges from land. But they were a long way from solving the problem. After years of fruitless effort by the conservation movement to alert the authorities to the dangers, official recognition stirred at Bremen, Germany in 1984, when eight nations bordering the North Sea signed a Declaration of Intent to reduce pollution. Plans were to be drawn up by the various states for discussion at a second meeting, which Britain hosted in November 1987.

The North Sea Forum, involving both voluntary and statutory bodies, was set up in 1985 to formulate Britain's policy. On the presentation of its report on 8 July 1987, it became the Marine Forum (subsequently the

Marine Forum for Environmental Issues) and continued its valuable work as an umbrella group for the interested parties. Its concerns revolved primarily around the lack of information available on the resources of the sea and the impact of man's activities upon it. Consequently, a much-needed management plan could not be drawn up. Equally, the effects of pollution could not be quantified. The British attitude was that pollution could continue until such time as it was shown to be causing damage. Things did not augur well for the second conference.

Fortunately, the European precautionary attitude prevailed and many recommendations to reduce pollution were made. The dumping of toxic waste was to be banned by January 1989, although material such as sewage sludge and shale, which were viewed officially as 'inert', would continue to be dumped. Incineration at sea was to be phased out by January 1995, by which time, too, the volume of nitrates and phosphates from run-off was to be halved. Plans were made to reduce pollution from offshore installations and the nuclear industry and to illegalise the dumping of ships' rubbish.

In 1990, the Third North Sea Conference promoted the idea of coastal zone management and introduced, for the first time, specific mention of

Plate 8.4 Pollution of the seas, abuse of the poorer beaches and excessive trampling of the better ones did not augur well for the littoral zone. Urgent action was needed both at local and international levels: sand dunes at Braunton Burrows, North Devon AONB. Courtesy of Countryside Commission

wildlife and habitats, particularly in seeking to protect dolphins and porpoises from speedboats and fishing nets. But most concern remained with marine pollution; sewage sludge and PCBs were singled out for more urgent reduction. Although not legally binding, the recommendations imposed a significant duty upon the North Sea nations to clean up their acts. Sadly, Britain tended to prevaricate again and the only stated intention was to end the dumping of sewage sludge by 1998!

As if to emphasise the moral aspect of the duty, nature itself began to speak up. A mysterious seal virus hit the headlines. Thought to be a form of distemper, it wiped out thousands of individuals, mainly common seals, within a matter of months. Experience abroad suggested that up to 80 per cent of the population would suffer. Seabird numbers began to plummet on the breeding grounds, notably in the Shetland Isles. There was a shortage of food due to overfishing and pollution was blamed. Other 'disasters' were noted around North Sea coasts. Meanwhile, the more regular – and ominous – signs continued; fish catches fell, sightings of porpoises, dolphins and other creatures declined, high concentrations of heavy metals continued to be found in shellfish. And in humans, sea-related food poisoning and infection, even hepatitis, persisted.

Industrial pollution, too, continued to be a source of embarrassment. As a result, the government integrated three inspectorates – for industrial air pollution, radiochemicals and hazardous waste – to form the Inspectorate of Pollution in April 1987. The new body also assumed responsibility for water pollution. Similar authorities covered Scotland and Northern Ireland. It was air pollution which caught public attention in the 1980s. The Swedes reported their watercourses affected by acid rain at the Stockholm Conference in 1972. The EC opened 380 monitoring centres and called for free access to all relevant information from member states. Britain remained strangely detached from the issue, despite allegations that she was a major 'exporter' of acid rain, until tell-tale signs in 1983 exposed its effects at home.

Acid rain, quite simply, forms when certain industrial gases react with the air. In the main, the worst pollutants are sulphur dioxide, nitrogen oxides and the hydrocarbons. Coal-fired power stations and car exhausts produce vast quantities of these. Not only is condensation made marginally more acidic, but also sunshine and visibility are reduced. Adverse effects in Britain, as elsewhere, rapidly multiplied. In the wetter uplands, fish populations declined in lakes and rivers. Birds, such as the dipper, were similarly affected. Trees and plants suffered; in particular, peatland flowers, such as bog rosemary, bog myrtle and sundew. But the signs were not confined to the uplands. Many native species of tree, such as oak, beech and yew, sickened in the south of England. And when the plants sicken, the animals begin to suffer. In retrospect, we now know that coniferous trees in the Pennines were affected by acid rain back in the 1930s. Kilvey Hill in the Lower Swansea Valley is a devastated graveyard of stricken trees, memorial to early industrialisation. The malaise had been with us for much longer than we knew.

When acid rain attracted European public concern in the 1980s, the first tentative moves were made towards a healthier solution. There was an International Acid Rain Year. Twenty nations agreed to reduce by 1993, their sulphur dioxide waste by 30 per cent of their 1980 figure. Britain was not amongst them. She stated that emissions had already been reduced by 42 per cent since 1970, by 25 per cent since 1980, and that their concentration in the air was one-third of the worst levels of the 1950s. Even though her emissions were still the highest in Europe, she smugly decided to sit this one out. The technology existed to reduce acid rain, but the will was absent and Europe's worst offender would only spend £600 million over ten years (1987–97) to reduce emissions by a mere 14 per cent of the 1987 figure.

In Britain, the main forum for the nature conservationists became Wildlife Link's Acid Rain Group, which faced an immense task of education in national cleanliness. So much remained to be done by Britain's industry to reduce its damaging effects. Industry, by its very nature, had often been an enemy of the environment. Originally, it affected only the land on which it stood. Then came the roads, the railways and the ports; the quarries, the mines and the spoil tips. Pollution entered the rivers and was swept out to sea. But with air pollution, the effects insinuated themselves throughout the whole environment. It was industry that first attracted bad press from the conservationists, long before farming, forestry or fishing. Consequently, industry reached out early to improve its image, generally using financial sponsorship as its tool. Cleaner techniques and less damaging processes come expensive and therefore often only in response to legal requirements. Even here, industry struggled to keep up and, in the face of a growing number of EC directives, the Confederation of British Industry had to set up an Environmental Legislation Committee. But much the best publicity derives from making high-profile, small-scale grants towards conservation.

One of the best known enterprises in this field is the Shell Better Britain Campaign, designed to give 'practical support to community initiatives in the environment'. The campaign grew out of the Better Britain Competitions (p. 161) and took on its new form in 1980. Several leading conservation groups became closely involved: the NCC, the BTCV and the Scottish Conservation Projects Trust, the Civic Trust and the Scottish Civic Trust, the CC and the CCS. Between 1980 and 1988, in addition to financing the campaign's administration, Shell UK committed almost £500,000 to schemes as varied as rubbish clearance and reserve management, building restoration and bat box distribution, an exhibition farm and island construction. The campaign offered specialist advice, not only on practical conservation but also on raising funds (p. 221).

Anticipating the increased enthusiasm of industry to support environmental causes, the movement founded the highly successful Conservation

Foundation in 1982. This became the main bridge between the two camps, marrying particular projects to specific sources of sponsorship from within industry and commerce. This tactic, to use the foundation's own hard-sell line, resulted in 'mutual marketing and PR advantage'. Widespread media coverage of its work 'alerted increasing numbers of organisations and companies to the benefits of commercial support for such enterprises both on a limited and global scale'. This high-geared marketing technique by the conservationists attracted an influx of funds to support a vast range of imaginative schemes. The *Conservation Annuals*, sadly no longer produced, catalogued the myriad different projects that were going on all over the country, often initiated by complete beginners. They showed concerned individuals reclaiming ponds and other forgotten corners, welcoming wildlife into towns by clearing footpaths and disused railway lines, by restoring cemeteries and installing bird-boxes. Children and pensioners, rich and poor were involved. Local and national, large and small, new initiatives and old chestnuts, the projects ranged beyond the normal concepts of conservation and sponsored great optimism for the future.

The BTCV developed an expertise in matching the funds of sponsorship to a pool of voluntary labour. In doing so, it achieved much remedial work on the environment and established itself as a major conservation force. By the middle of the decade, over 500 local groups and 1,000 schools were affiliated, along with 8,000 individual members. In 1986, nearly a quarter of a million work-days were put in by over 45,000 volunteers on 3,000 sites in England, Wales and Northern Ireland. A sister organisation, the Scottish Conservation Projects Trust, had been formed in 1984 as a natural progression of BTCV's Scottish region. Great emphasis is placed on training in skills such as drystone walling, hedging and fencing, foot-path construction and the correct use of tools. Equally valuable tasks concentrate upon the clearance of ponds, paths and litter. Guidance is available to the many corps which have inevitably arisen within local trusts for nature conservation in the wake of BTCV's own success.

THE WORLD CONSERVATION STRATEGY

Of all the bridge-building exercises, the most vital was to span the gap beyond which stood the economists. All efforts to prevail upon farmer and forester, industrialist and politician were made against the flow of economics. While all those interests at least found it possible to pay lip-service to conservation, economic theory did not recognise it at all. Mineral resources and natural food increased in cost as they became worked out, until it was uneconomic to exploit what remained. But land-scape and wildlife did not exist in economic theory; they could be altered or used up at no monetary cost. The conservationists had come to realise, at last, that this was not the case. They had to shelve the ethical and spiritual arguments upon which they had relied for so long – and which had never cut any ice in the world of economics – and find a value for their cause which the economists could measure. The nature conservationists only came to realise this when they encountered the down-to-earth policies of the new conservationists. This, I believe, is the most important result of their coming together as far as the nature conservationists are concerned. It was undoubtedly with much relief that some pure naturalists began to understand that there was indeed a good economic case for conserving the environment.

The first coherent statement was promoted by the international movement as the World Conservation Strategy (WCS), instigated by IUCN and UNEP. Launched in the fear that we were living beyond our means in terms of natural resources and wildlife and setting ourselves apart from the ecosystem upon which we rely, the strategy was very wide-ranging in content and global in concept. Knowing that vested interests in short-term profiteering would condemn it as the return route to peasantry, its authors emphasised the precept that conservation is the key to continued prosperity; that conservation and development are mutually dependent and not mutually exclusive.

The WCS, three years in the making, was launched in thirty-four countries on 5 March 1980. Subtitled 'Living Resource Conservation for Sustainable Development', it was heralded in the British Parliament as:

> a realistic and unemotive restatement of the evidence that conservation of our natural and living resources is essential to the economic and social welfare of society and is entirely compatible with sustainable development.

In his *Overview* of the strategy, Johnson (1983) summarised it thus:

> The World Conservation Strategy . . . showed that over-exploitation of resources, loss of genetic diversity and damage to ecological processes and life-support systems have dangerously reduced the planet's capacity to support people in both developed and developing countries. It sought a new partnership between conservation and development, to meet human needs now without jeopardising the future, and called upon each country to prepare a national conservation strategy tailored to its own particular problems and characteristic culture and economic conditions in order to achieve this.

Inevitably it met with a very mixed reaction amongst the conservationists. Moore (1987) saw it as 'a chance, perhaps the last chance to achieve effective conservation on a world basis'. Nicholson (1987), in the minority, felt that it did not go far enough: 'Where it most conspicuously failed was in omitting to tackle the grave implications of grossly excessive human population increases upon the health of the biosphere'. Most people, including the converted, feared that its implications were too far-reaching

Plate 8.5 The WCS sought a new partnership between conservation and development, recognising that the former was essential to economic and social welfare and entirely compatible with sustainable development. Photograph by M. Charity. Courtesy of Countryside Commission

for it to gain acceptance; that its message would seem 'on some lips a cynical and trivial, or blindly idealistic one' (Adams 1986). But, for the conservationists who thought beyond their own narrow realm of interest towards the inevitable conclusions of their beliefs, the WCS merely stated the theory that had underpinned their ethic since its earliest days and endeavoured to harmonise it with other codes of living.

The sceptics will say that the Strategy is yet another high-sounding international declaration launched in a blaze of publicity but not followed up and quickly forgotten along with so many of its predecessors. This one should be different. It is based on three decades of global work, and has been agreed internationally and is endorsed by agencies such as UNESCO and FAO. It is also the first real attempt to crystalise competing interests into a single unified objective.

(Dawson 1980)

But Dawson's optimism was not based upon the level of official enthusiasm within the UK. 'So what has been the reaction of the British Government?' he asked in 1981.

After the enthusiastic welcome by Mr Michael Heseltine, the Secretary of State for the Environment, everything has gone quiet. In other countries work has been started on national strategies, but the absence of any official interest has prompted the voluntary bodies to combine in a joint project which should lead to a UK document for presentation at a conference sometime in late 1982.

Those voluntary bodies were the WWF(UK), which provided the finance, both Countryside Commissions, the Royal Society of Arts, CoEnCo and the official NCC, all working under the UK committee of IUCN. The resultant national strategy, entitled *The Conservation and Development Programme for the UK* consisted of Johnson's *Overview*, published in 1983, and seven detailed reports which followed in 1984. These looked at:

- industry,
- urban environment,
- rural environment,
- marine and coastal environment,
- UK overseas environmental policy,
- environmental ethics and conservation action,
- environmental education and sustainable development.

Notwithstanding the fact that their government had not assisted in any way with drawing up the national strategy – and that many leading ministers remained blissfully unaware of the WCS – the conservationists still held high hopes for the British response.

The inability to develop a broader kind of nature conservation [since 1949] should be regarded as the nation's failure; the opportunity was there but it was not taken. The Conservation and Development Programme for the UK has taken up the challenge again with great vigour and well deserves to succeed but the resistances are still considerable.

(NCC 1984)

They were indeed. By 1986, with twenty-nine developing and eleven developed countries working on their national strategies, Adams could report of Britain: 'The whole exercise was largely ignored by the media, and to date little interest has been shown by government'. The only practical response by then had been the creation in July 1984 of the UK Centre for Economic and Environmental Development (CEED). This was an advisory non-governmental charity supported largely by donations from industry, environmental bodies and charitable trusts. Its overriding aim, in line with the WCS, was to reconcile economic progress with conservationist thinking, mainly through publications and seminars. From its own brochure: 'With the growth of environmentalism as a dynamic political and moral philosophy throughout the world, conflict between conservation and development interests is frequent and sometimes damaging. CEED's aim is to replace conflict with productive partnership'.

Having had a hand in setting up CEED, the British government apparently felt that its part in the WCS and the national response was complete. Nor was there anyone clamouring to tell the politicians that they were wrong. The strategy had gone over the heads of many conservationists, and the economists remained clearly unconvinced. This 'high-sounding declaration' was no different to any of the others; in less than five years, it was forgotten in Britain. But there was a glimmer of hope left, thanks in no way to the ardent British conservationist but to the perspicacity of the UN. This body set up an independent World Commission on Environment and Development, chaired by the Norwegian Prime Minister, Mrs Gro Harlem Brundtland. The commission studied a range of relevant topics, from pollution and wildlife conservation to energy and food security. It concluded that nations across the world must commit themselves to sustainable development to ensure a just, secure and prosperous future for all people. When the official Brundtland report, *Our Common Future*, was launched in London in April 1987, it came as a timely reminder of the principles of the WCS. At about the same time, the EC was embarking upon its fourth action programme which sought to make protection of the environment an integral part of economic and social policies. Two months later, the Council of Europe was calling for a European Conservation Strategy. It seemed, perhaps, as though the WCS was not going to be forgotten quite so easily on the international scene.

One concrete result of the Brundtland commission was the World Heritage Convention, which recognises both natural and cultural properties with 'outstanding universal value' and, realising international obligations to help preserve them, runs a World Heritage Fund. Acceptance of sites for designation is proving very difficult. For nominations under the 'natural' category, IUCN submits a report to an international committee. In 1995, thirteen British sites had been accepted, of which only the island of St Kilda and the Giants' Causeway might be classed as 'natural', but some ground was lost at the outset by the failure of the UK to ratify the convention until 1984. The Lake District failed in a bid for acceptance as a 'natural' site in 1987, and renewed efforts relied upon its 'cultural' value. Pressure is also being applied to have the Cairngorms recognised.

Given its reluctance to acknowledge a convention on an idea as familiar as conservation sites, it is no surprise that Britain, along with many other nations, has persistently ignored most of the tenets of the WCS. Moreover, she has hindered some European moves to accept its principles. In 1987, EC funds for research into environmental protection were exclusively directed towards pollution or bird projects, because those aspects were governed by existing directives. Other fauna, flora and habitats were not, and therefore failed to win any funding. To alter this anomaly, a Fauna, Flora and Habitats Directive was proposed in 1988, seeking to safeguard areas of particular value to wildlife. Britain was one of the two member states not to support it, either missing the point entirely by taking the view that 'we do enough already' or shying away from the responsibility to protect habitat types, such as hedgerows and ancient woodland. This negative response provoked uproar from the conservationists who saw great promise in the directive for wildlife protection not only at home but in other, more backward European states (p. 240).

That such a basic principle as habitat protection should meet with such resistance in official circles is symptomatic of the conservationists' struggle. The wider implications of the WCS never got off the ground. It is a measure of its failure that many conservationists remained ignorant of the strategy's existence. Economists and politicians were no more aware of it and its message was not communicated to the British public at large.

But before any of these strategies – global, continental or national – are written off as failures, we must stop to consider the reality of what they stood for. To suggest that conservation and development could go hand in hand was stunningly novel to many people in 1980, most conservationists included. To attempt to integrate the two had all the hallmarks of a pipe dream. A sustainable society which satisfied the physical and mental needs of all people seemed further off than the furthest planet in the light of city riots, personal degradation, social inequalities and moral collapse. To seek a stable and sustainable economy might be a pleasant aim, but the strategy also demanded that new technologies be found to replace contemporary resource-greedy and polluting methods; that non-renewable resources be saved; that renewable supplies and infinite energy sources be used; in short, that demands of all sorts be reduced and that trade and industry become adaptable to the changes ahead. Socially, increased personal knowledge and involvement in local community and landscape were expected to magic up a sense of identity and belonging in every individual and thereby create an atmosphere of concern for the environment. Barriers were to be swept aside, whether they were educational, ideological, institutional or financial. Such was the scope of the World Conservation Strategy.

When one considers the fundamental changes that were demanded in technology, economics, policy- and decision-making – the whole spectrum of human activity – and the mammoth re-appraisal of attitudes required, it was obviously decades too early to say whether the WCS had failed or not. But the committed conservationists, who welcomed the strategy with missionary zeal, might be forgiven for fretting. Much of what it contained echoed their long-held hopes and fears; the arguments were not so novel for them nor the required changes so great. Perhaps they simply expected too much too soon.

Or was it that the conservationists were being narrow-sighted and parochial again? Much of the subject matter of the strategy fell outside the field of nature conservation. True, it considered wild animals and plants as resources of genetic diversity, important for the maintenance of essential ecological processes. As such, they were to be conserved in all environments, with special reservoirs in the form of nature reserves. The conservation of the resources upon which they depend, and the control of pollution and damage, were seen as vital aspects in their management. This, of course, was old hat to the nature conservationists who, by definition, always used these arguments. But perhaps they were closing their eyes to the greater part of the strategy. Perhaps they did not consider the fundamental changes facing so many aspects of society. It would not have been the first time that nature conservationists had failed to see the problem from the other side. Even the new conservationists could hardly have ignored the wider implications of the strategy, unless by dint of wilful blindness. Such monumental efforts at bridge-building deserved better than that.

There were elements of the strategy which held immediate promise for nature conservation and recreation, if only they could gain acceptance. One of its overriding precepts was that consideration had to be given to all aspects of a planned activity before a decision was made on whether to go ahead or to abandon it. Thus monetary worth was but one factor; its usefulness to people or society in general must also be considered. Unbiased regard for its intrinsic value was required. Some attempt was needed to assess its symbolic worth, that most abstract of qualities which is so difficult to quantify and yet so meaningful to the human spirit. It may have seemed quixotic to aim so high, but it was surely necessary now that human activities were affecting all parts of the planet to such a great extent. Only this full consideration could hope to result in the wisest decision, either to conserve or develop, or to take a middle course. Nobody could complain if such painstaking homework had been done.

To that end, the WCS called for two types of assessment: one of the area in which a change of land use was planned and one of the likely effects of the proposal itself. The first was the Ecosystem Evaluation (EE). This seeks, ideally, to survey all parts of the globe and to assess their relative value to various activities – agriculture, nature conservation, urban use, recreation – any type of land use whatsoever. Furthermore, these categories can be subdivided. Thus, in terms of nature conservation, an area might be of particular value as a habitat type; as a reservoir of genetic diversity, or as a haven for threatened species. The EE seeks to quantify the cultural and biological as well as the economic importance of an area, so that we learn the true value of the place we plan to alter. The Environmental Impact Assessment (EIA) examines all the likely effects of any proposal and its alternatives, whether those effects are perceived to be good or bad. These issues include impact on wildlife and landscape, pollution and noise, alongside the more traditional planning considerations. The two assessments combined can provide the decision-makers with a well-informed base from which to work in the best long-term interests of the planet and the human race. How else could we ever have expected them to get it right?

Not surprisingly, the newer concepts were slow to gain acceptance in official circles. Ecosystem Evaluation seems as far off as ever in Britain. It remains a matter for the conservationists to keep under review all the valuable wildlife sites so that meaningful assessments can be communicated to the planners and politicians when development threats are foreseen. That indeed is a prodigious task for a predominantly voluntary movement. Furthermore, who will carry out unbiased EIAs? The EC eventually issued a Directive on Environmental Assessment of Development Projects in 1985. It gave member states until 3 July 1988 to implement planning procedures that encompass EIAs. Britain got around to introducing them for industrial developments in 1990, just as the EC was planning to extend them to major agricultural projects. We have yet to make up the lost ground.

CAMPAIGNS OF CONSERVATION

It was one of the precepts of the WCS that, for conservation to become an integral part of all human activity, levels of awareness of its principles must be raised. So, while the leading lights were thrashing out world strategies, the sterling work at grass-roots level was recognised as no less important a part of the master plan. The 1980s saw a number of traditional campaigns, and the decade became a maze of designated days, weeks and years. The Council of Europe ran its Water's Edge Campaign in 1983–4 and the WWF and IUCN had an ongoing Wetlands Campaign. Fifth June 1984 was World Environment Day. The Civic Trust ran the first of its annual Environment Weeks in 1985. Originally an amenity campaign, it widened to encompass nature conservation in 1988 when it was adopted by the RSNC, the BTCV and other conservation bodies.

The RSNC launched its British Wildlife Appeal in October 1985, with the slogan 'Tomorrow Is Too Late'. It aimed to raise £10 million in five years, half the sum through a national appeal and half on a local scale through the county trusts. Of this, 40 per cent would be spent on the purchase of land for endangered species or threatened habitats, 40 per cent on land management and 20 per cent on promoting greater public

---

New societies for the 1980s included

1982	Scottish Wild Land Group
	British Hedgehog Preservation Society
1983	British Dragonfly Society
1984	International Mire Conservation Group
1986	National Federation of Badger Groups
1987	Whale and Dolphin Conservation Society
	European Cetacean Society
	Living Earth
1988	Plantlife

---

awareness of the conservation cause, especially by introducing youngsters directly to our natural heritage. By the society's seventy-fifth anniversary, 1 March 1987, £4 million had been raised. Another of its initiatives was the National Wildflower Week each May, which aimed to increase concern for this neglected aspect of our wildlife.

National Bat Year was launched on 25 February 1986, the year that saw the first legal action taken under the Wildlife and Countryside Act against the unlawful killing of bats. The FFPS was funding a bat conservation officer to co-ordinate the work of the growing number of bat groups. This progressed to the formation of the Bat Conservation Trust in 1991. Four years later, it could boast over ninety local groups and two thousand members. The FFPS also ran a 'Save Our Snakes' campaign on behalf of this other non-favourite.

The year 1987–8 was the Council of Europe's Year of the Countryside. Under the slogan 'Let's make the most of our countryside', it aimed to reverse damaging trends in the rural environment, particularly rural decline, urban sprawl and 'the growing imbalance between man and nature'.

For a year starting on 21 March 1987, the EC promoted European Year of the Environment (EYE). In Britain, the three main themes were the conservation of the natural environment, the control of pollution and waste, and the improvement of the urban environment. The inaugural event in Britain was the planting, within Sherwood Forest, of a 24 ha oakwood map of the country. Every primary school in the UK was represented by one of the twenty thousand saplings. During the year, awareness campaigns were mounted on various topics, such as forests and clean beaches, and there were 'weeks' for the National Parks, wildflowers, water management, tree planting, acid rain and others. Many events were co-ordinated to run throughout Europe, but very many more initiatives were organised at local level.

And so the list of campaigns, years and weeks continued to grow. But how do we evaluate the success or otherwise of an awareness campaign? Concrete results are unfathomable; that practical project might have happened regardless, those new members might have joined the society anyway. Shifts in attitudes are even more inscrutable. History, the benefit of hindsight, is the most accurate tool. It is clear that EYE was far less effective than ECY. For, in 1970, the message of conservation was still new to many people. Awareness campaigns of that sort were still novel for the movement. By 1987, everyone knew that there was concern for the environment even if they did not share it. National Wildflower Week had already been planned; many other EYE initiatives would have been launched without the official EYE campaign. So how effective was it? For how much longer might such campaigns remain meaningful?

European Nature Conservation Year (ENCY) 1995, co-ordinated by the Council of Europe, planned to enhance public awareness of, and conservation in the wider countryside away from designated areas. Unless you were directly involved in it, you might be forgiven for missing it altogether; it was probably the least successful of the environmental years so far. Were we, then, promoting too many designated weeks and years?

A similar question, relating to the number of societies, had vexed the movement since the 1970s. Were there too many groups? Should not those with similar interests come together for greater strength? By doing so, would they lose members and influence, or gain them through their broadened base?

The movement valued its wide spectrum of individual societies, for many of which growth had been impressive (Table 8.2). The fear that membership would fall off scotched moves towards merger time and time again. Each body jealously guarded its status and its niche. It valued the members who supported its particular aims and aspirations, and it looked to the various umbrella groups to co-ordinate activities within the wider movement.

Table 8.2 Membership of various voluntary bodies (in thousands)

Organisation	1971	1981	1990	1995
NT	278	1,046	2,032	2,323
NTS	37	110	218	234
RSPB	98	441	844	890
RSNC	64	143	250	252
WWF	12	60	247	200
FoE (England & Wales)	1	18	110	200
CPRE	21	29	44	45
Woodland Trust	—	20	66	62

Source: Brown 1992 and individual organisations

But this independence has often been taken as a mark of the movement's disunity, both by those within it and by those without. This is something that the movement can well do without, dogged, as it is, by genuine topics of controversy, such as bloodsports and public access.

The main value of awareness campaigns and umbrella groups has been to promote the image of a united movement. By co-ordinating local activities and smaller-scale initiatives, by making the total effect greater than the sum of its parts, they highlight the strengths and potential of the various elements standing together. As the field of interest continues to widen, stronger ties will be required to bind together a movement which will come to include the organic farmers, the green consumers, the advocates of village life, the champions of full employment. The spectrum of involvement should be vast. Awareness campaigns have moved on from being a means of promulgating the plight to become an advertisement for the efforts being made to combat it. EYE probably instigated very few initiatives of its own, but it did provide a focus, a common theme for activities as diverse as birdwatching and the improvement of industrial technology, a painting exhibition and an opinion poll on clean beaches, moves for safer cycling and the recycling of waste. It presented the conservation movement as a broadly-based force to be reckoned with. It did not need to spread the word that we must conserve; its success was in integrating so many themes based upon the conservationist ethic. People in just about every field of human activity ought to have been moved to review their attitudes and their practices.

Like wheels within wheels, individual campaigns such as National Wildflower Week have the same purpose. They are an umbrella to cover a variety of activities, a focus for interest; a central theme to attract greater attention than might the individual projects.

The conservationists needed mechanisms like these to hide some of the blatant disunity that was dogging the movement, if we can call it such, even at the end of one hundred years. Limited visions of their role had always prevented them from coming together, either amongst themselves or with other countryside interests.

> The case for conservation is weakened by lack of coordination between those concerned with scenery, wildlife, antiquities, and freedom. The arguments, naturally, differ, but the objectives are often the same. One proponent often accepts compromises which weaken the case of the others.
>
> (Rackham 1986)

Nature is indivisible and, in our shrinking world, even the most selfish conservationist must take a global view. We saw in Chapter 2 how a divorce from nature caused a spiritual void which gave impetus to the infant movement. The distance between humankind and nature which now drives the wider 'green' movement has little to do with spatial distance and more to do with the methods we operate to destroy the natural world. What we can now do has driven us further than ever from an understanding of nature. But there are fewer people now who profess to believe that we can control any effect we may have on the environment. The timely revelations of truly global threats which respect no social nor national bounds have pushed the environment into the limelight. The nature conservationists are now but part of a widening movement which must carry forward the cause of the environment. They cannot disassociate themselves from the whole cause, just because its basis is different; because it exceeds their self-imposed limits of concern for animals and plants. Yet, astonishingly, it seems that they may try to do so. Blunden and Curry (1990) say of the RSNC: 'They, just as European governments are likely to, see the global environment and the countryside as quite distinct'. They fear that nature conservation will drown in the new debates on global issues. Drown it will if it does not promote itself as an integral part of those issues.

The new cause is no more selfish than their own has always been. Each and every environmental issue ultimately concerns every conservationist, whatever his or her personal basis of belief. Pesticides or pollution perhaps left some part of our heritage untouched; atmospheric decay will not. Nature conservationists must now link arms with every other type of environmentalist and together they must seek that clarity and consistency of outlook that has eluded them in the past. Now that it has attracted close public scrutiny, the movement must not fail to impress.

It was the combination of the words and the music which enabled the environmental movement to take off and to gain an impetus that, to the astonishment of many outside its ranks, it has never lost. If that is to remain the case, it must not be taken for granted; in particular all those responsible for running the movement, in its happily loose and decentralised form need to understand what makes it tick.

(Nicholson 1987)

The movement may well remain decentralised and so profit from its broad base, but it must work for higher levels of communication, co-ordination and co-operation. In deciding its aims and priorities, it must consider all problems with an open mind: nuclear power, the local pond, barrage schemes, trading practices, blood sports, microspecies, the tropical rain-forests, organic farming – everything in the environment. It is clear that we cannot have everything; choices have to be made.

what the environmental movement should have tried to do is to draw up a comprehensive, realistic and persuasive programme of its own for the future ... once the scale of the development pressure had become apparent. This would have taken thought, conferences, effort and imagination. But it would also have taken something else – the willingness to face up to unwelcome choices.

(Shoard 1988)

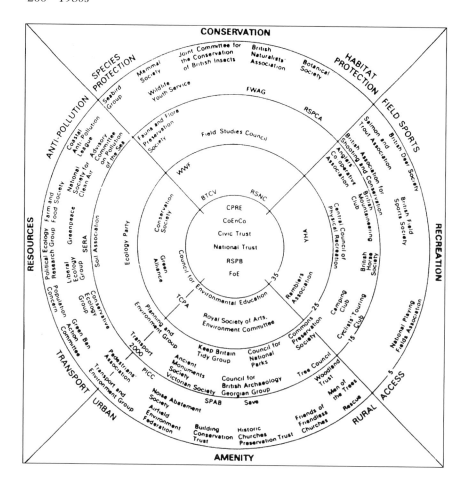

Figure 8.1 The organisation of the environmental lobby (from Lowe and Goyder 1983: 81). Each group was asked with which other groups it was in regular contact. Groups have been allocated to different bands according to the number of contacts attributed to them (rather than claimed by them). Thus, groups in the outer band are in regular contact with five or more other groups, but less than fifteen. By and large, groups tend to be in contact with other groups in the same sector or with groups in the central core. By courtesy of Unwin Hyman Ltd.

No longer can it be permissible for a group or society to hide behind its charter, saying, 'These things are not for us to consider'. These things concern every conservationist and affect the entire environment in the long run. To shun the responsibility of discussing them is to abdicate the role of the conservationist.

The conservation bodies should achieve closer integration, defining a collective approach, at both policy and operation levels, and distributing individual efforts according to respective strengths and responsibilities. ... Common objectives and means of establishing more mutually supporting relationships should be identified.

(NCC 1984)

Compromise will have to be made but the experience will be valuable. Many conservationists forget that their own reasons are different from those of their colleagues. What hope then do they have of appealing to those outside their ranks, to those who have no comprehension of the need to conserve? They have a very different view of what the countryside should be; of what they would expect to find there and what pleasures they would hope to derive from it. Who can say what landscapes they should find exciting or relaxing? Perhaps the countryside holds no conscious importance whatsoever for a few. It is a problem which is growing fast for the amenity wing of the movement. The countryside leisure industry is in danger of self-destruction. People in the countryside damage what most of them come to enjoy; they clutter the scenery, upset the wildlife, trample the plants, disturb the peace and destroy the isolation. Development of tourist facilities and paraphernalia threaten the very existence of the countryside. In deciding its priorities, the amenity lobby must not presume to decree the type of landscape that we should wish to see. Through consultation, it must provide the whole spectrum of requirements.

The nature conservationists likewise are in danger of upsetting other countryside users. They have always tended to seal off areas for the sole use of wildlife. Compromise must be made here, too. Countryside can be valuable for its landscape as well as its habitat.

I think that there can be no doubt that awareness of the need for conservation of nature, active interest in the idea of conservation, will have developed greatly by the end of the century. But it will still be a minority interest. We delude ourselves if we think that it will not.

(Vesey-Fitzgerald 1969)

Some conservationists do continue to delude themselves. Realising its vital importance to everybody and everything, they blind themselves to the fact that most people remain disinterested and unmoved. Thus the need for a message of universal appeal – the future of mankind itself – preached with guts and determination by a united and credible movement. While conservation for its own sake will remain a minority interest, conservation for the future of mankind must come to concern everybody, so that:

people will want to follow the environmental state of play in the way that today they follow the latest trade figures, financial indices and the fortunes and form of football clubs and athletes.

(Moore 1987)

Moore looked forward to a day when conservation would surpass party politics – which it had only just entered – to become a national objective, alongside peace and health. Indeed, conservationist ethics can go a long way towards ensuring peace and health, and our ultimate aim must be to see them realised not on a national, but on a global scale.

UNIVERSALITY

> It is one of the unfortunate paradoxes that a conscious sense of conservation seems only to come after a long period of devastation has made the need apparent. We ought to be able to do better than that . . . . Time is not on our side and our present enlightenment may not go far enough.
>
> (Darling 1971)

It is all very well to expect great things of a wide-based and unified movement, but unity by itself is not enough. In order to convince the majority of the relevance of its minority interest, the movement must escape its fringe-body image and promote itself as the fundamental basis of our mainstream existence. 'Amenity is an extra, something which . . . can be added to life on top of the real business of existence' (p. 6). It retains the flavour of a dressing, a pleasant topping to the menu of life but one which can be left off if necessary, if the finances do not allow. Consequently, the movement has been seen as part of an alternative lifestyle.

Unfortunately, circumstances have tended to work against the movement's acceptance as an attractive proposition. Quite apart from its own mistakes and shortcomings, it has generally been forced to oppose, to counter and to protest. The very nature of its task has been to prevent real, imagined and prospective wrongs. It has had to seek safeguards, call for contingency plans and foresee calamities that have not always (yet) come to pass. In demanding recompense and pursuing redress, it has seemed to oppose all progress. Its limited resources have prevented it from applauding very warmly those steps taken in the right direction; it has always needed to turn quickly to tackling the next task. This negative stance, though necessary, has not been its most attractive feature. Much of its limelight has been won in the wake of environmental disasters, which make for a good press story but quickly become relegated out of print. The conservationists just have time to say, 'We told you so,' and 'You should have,' before the limelight is extinguished and they are left again looking negative, out of touch with reality and wise after the event. Their other campaigns can seem trivial to the point of pettiness, perhaps because they appear to be fought on a localised scale or on behalf of only one or a handful of species. Here, conservationists suffer from their failure to teach the fundamental point of their ethic: that every organism is a vital link in the web of life. Most people, failing to comprehend the importance of the conservationist standpoint, find it easier to view it as an alternative lifestyle, which pursues petty and irrelevant goals.

Much of this is attributable to those scientists and conservationists who persist in keeping their heads in the clouds. High-faluting statements couched in specialist terms will register only with the specialist, however much the argument may relate to other people. Unrealistic claims serve to weaken, not to endorse a case. The Doomsday era of scaremongering had to mature into an age of balanced discussion well founded upon firm evidence.

Armed with the facts, the conservationists would have a solid base for their campaigns. But their greater challenge has always been to put across the message in a meaningful way.

> In fact, the principal lesson we learnt from experience was that communication with the authorities and the general public was crucial and was much more difficult to achieve than doing the research . . . . The failure of society to act shows that conservationists are failing to communicate effectively.
>
> (Moore 1987)

For there are others who remain unrealistic and so do the cause no good. Optimism there must be, but its goals must not lie beyond our reach. There are still those, the preservationists, who would leave nature to its own devices. They are out of touch, not only with our place on earth but also with the reality of their own cause, for so little of our landscape and habitat is natural and, without active management, would deteriorate in value. It is inevitable that, having moulded so much of the environment, we must now continue to manage it. In the past, 'the very word "management" as applied to nature was strange and unpalatable, smacking of a contradiction in terms' (Nicholson 1987). That is just one of the attitudes that must now be finally stamped out.

> People who care about nature tend to be romantics and prefer nature untamed. The very act of categorising the wilderness is repellent, especially if done by a government department. I can sympathise with this attitude, but it is totally unrealistic in the world of today. Nature is under serious threat and if we are to conserve it effectively, we must solve the problems connected with it through the ordinary work day channels of finance, legislation and politics.
>
> (Moore 1987)

Then there are those who will not compromise; those who feel that their cause should be embraced without question by all or that it can thrive in isolation from other land uses. If the tendencies to be too scientific or unrealistic seem to be waning, there is still a danger that the unwillingness to compromise may even increase.

No one could deny that the acceptance of compromise is essential to the functioning of a democratic society. But in nature conservation the great compromise has been made already, through the surrender of so much of our heritage of nature to development for the national good. Since 1949, when nature conservation first existed in organised form, it has been compromising all the time, often indeed with very little choice, by paying regard to the needs of other land-use interests. In some parts of Britain there is little if anything to compromise about, and we feel that the time has come for other interests to pay regard to the needs of nature conservation.

(NCC 1984)

The exasperation is clear. Almost apologetic in the past, the conservationists were being forced, by an increasing sense of urgency, to make such comments. But the days of scaremongering should have passed. Just as preservation is no substitute for conservation, so entrenchment is no substitute for compromise. To remain convincing, the movement must stay calm, collected and unemotional. It will have to accept compromise.

Many commentators remain sceptical of this approach. O'Riordan (1989) felt that 'partnership can shackle good intentions and bog down the activist' and lead to a situation in which the decision-makers tend to seek the middle ground – which, in most cases, represents 'a serious erosion of conservation values'. Then he states that there is no distinctive leadership because the movement is so dispersed, and that it is not putting across 'just what dire straits nature conservation is in'. This is a contradiction. Closer partnerships help to emphasise the conservationist viewpoint amongst those who fail to comprehend it. Closer ties within the movement make it look more genuine. Professor Robert Worcester, chairman of MORI, in a 1994 poll for Friends of the Earth, commented that 'many of the organisations appear to compete, not complement' and gave this as one reason why environmental concerns had lost their edge since 1990. The leaders which the movement so badly needs at this time will be those who succeed in bringing together a more unified movement to work with the disinterested and the non-believers.

Compromise, though, does not equate with weakness or defensiveness. Compromise can be a powerful force. Much interest and invitation from other fields has recently become evident and it should extend itself rapidly as the conservationists advertise their relevance to all walks of life. Yet this outside concern seems to have taken them by surprise. They will lose it if they do not act warmly and positively. They must show other interests their willingness to meet them half-way, or else convince them of the need to travel further down the conservation trail. They must show people that they, as individuals, can make a difference. Human nature tends to shy away from the problematic and the misunderstood; from apparently insurmountable odds. The new cause of conservation has been one such enigma to most people. They have tended to put up barriers and to utter that it cannot be helped. Times change, progress must be made and we shall have to accept the outcome. The movement must hold out its hands to these people, must show them how valuable they are, each and every one, to the conservationist cause.

To this end, the movement has had to reconsider its tactics. The short-term, specific and localised campaigns need to continue, of course, in spite of their negative aura. But the overall image must become one of positive action involving all people and interests to find long-term solutions to man-made problems for the environment. The conservationists must become politically and socially astute. They must be versatile, agile and quick-witted. They must be opportunist. But most of all, they must be convincing. Their cause will then move from being a fringe activity within an alternative lifestyle to becoming the mainstream of all human activity.

> What is needed are integrated policies, not attempts to meld incom-patible policies. That way lies conflict and – to be realistic – defeat for conservation. This is, of course, the message of the World Conservation Strategy.
>
> (Adams 1986)

The WCS placed great importance on influencing policies as they are drawn up rather than reacting to them retrospectively. It proposed a sustained drive to educate government, business and the professions by explaining the relevance of conservation to their lives. It saw the need to make conservation a part of everybody's existence. The reasons could be as selfish or as personal as they liked . . . as long as conservation mattered. In fact the strategy was the movement's best effort yet to pull together all the various threads.

If we accept that it was the product of a unified movement, as it sought to appear, why then did it lack credibility amongst conservationists, let alone awareness amongst the wider public? The answer must be that the movement still failed to sound convincing, even to its own supporters. It just was not plugging its policies and its new positive image strongly enough. Incredibly, it seemed not to realise just what powerful interests and inertia it was up against. Modern social values revolve around material wealth; a prominent aspect of human nature is greed. Combined, they spawned a society of consumerism based upon the economic creed. But conservation does not envisage continuing economic growth; it is the antithesis of consumerism. In seeking to limit material wealth to an accept-able and sustainable level, it plans to alter the whole basis of the human social ethic; the very essence of our time on this earth. That goal is so awesome and pandemic that it explains why many so-called conserva-tionists remain unconvinced of their high ideals. Until such time as they accept them fully, they will continue to lack credibility, and efforts such as the WCS will fail through fragmented support.

The 1980s saw the movement wake up to its duty to involve and educate those with other aims and interests in life. This was, in no small part, due to the fact that many of the decade's initiatives had failed: the Wildlife and Countryside Act, the World Conservation Strategy, European Year of the Environment. As a result, efforts were made towards building some bridges and there were a few promising achievements. But the task remained a formidable one and the speed with which it might be achieved depended largely upon the ability of the conservationists to spread their message convincingly; to present themselves as a positive and forward-looking force, always fresh, meaningful and firmly on the ground; to shake off the aura of negativity and exclusive insularity; to promote conservation, not as the alternative way of life, but as the only feasible way of life.

9 1990s: Government inertia and global initiative

THE NATURE OF POLITICS AND THE POLITICS OF NATURE

> Government holds the scales, and ultimately it will have to decide where it wishes the balance between nature conservation and other sectional interests to lie. And the view taken by government, in representing the people, will depend on its perception of public concern.
>
> (NCC 1984)

Politics has a lot more to do with words than actions. In a democracy, government puts in place the machinery to run the country as the majority chooses. If a mechanism is sufficiently unpopular, it will fall into disuse or disrepute and be dismantled or replaced. If enough people call for additions or refinements to the machinery, they will follow only if the case has been put forcibly enough to overcome the inevitable inertia of the *status quo*.

The fact that British politics still viewed nature conservation as a minority interest was painfully clear.

> whatever the torrent of printed and spoken words what is lacking is not so much in the legislation as in the legislators. The real determination, dedication and will to save wild country appears to be present in only a relatively few Members of Parliament, and even fewer Ministers for the Environment have yet demonstrated an active recognition of the value of our last wildernesses.
>
> (Sylvia Sayer, in Lowenthal and Binney 1981)

In 1982, two years after the launch of the WCS, both the Foreign Secretary and the Minister for Overseas Aid publicly admitted that they knew nothing of it. The UK government's response was eventually trumpeted, six years after the event, in a glossy booklet: *Conservation and Development: the British Approach*. It was blatant, empty propaganda, a sales brochure from the world of politics that carefully avoided neglected issues and areas of controversy. It made the sums of money spent on conservation sound generous. It made the various protective designations sound meaningful, as if a barrage could never be built across the Cardiff

Bay SSSI or a motorway through Twyford Down AONB. It carried a half-page photograph of the *Torrey Canyon*, of all things, as if to suggest that the *MV Braer* or the *Sea Empress* could never run aground in the future (p. 219). It showed Leighton Moss in Lancashire with the caption: 'an RSPB nature reserve designated as a Wetland under the Ramsar Convention and a Special Protection Area under the EC Directive on the Conservation of Wild Birds'. You might have been forgiven for thinking that government had expended much money and effort here. Of course, the RSPB bought and managed the reserve through voluntary contributions. Official designations had cost the government nothing. The truth is that designations of Ramsar sites and SPAs was very tardy indeed and guaranteed nothing for the future. Of a possible 238 sites identified by 1992, only forty-seven had been officially recognised, and twenty-seven of them, along with 117 of the others, were threatened with development, mismanagement or neglect. In 1990, Abernethy Forest was granted SPA status. Would this have been the case had the RSPB not just spent £2.5 million purchasing it from private ownership? Of this booklet, Nicholson (1987) states:

> It shows recognition of the growing power of the environmental movement, and readiness to respond to it and work with it. Implicitly it confirms that the government will go along with the movement to the full extent that the movement can prove its case to public opinion, but no faster.

That is the nature of politics and, until such time as the conservationists could instil their message in the majority of British people, the government could not be expected to take more positive action.

Nevertheless, the conservationists had shown a willingness to foster new links with the politicians, realising how vital central policy decisions were to their cause.

> Ultimately, it is clear, we shall lose the whole game if governments cannot be persuaded to take on most of the burden.
>
> (Fitter and Scott 1978)

So while the protracted task of convincing the public went on relentlessly, the movement had also stepped up its direct approaches to the world of politics. The Green Alliance was formed in October 1978 to alert ministers to conservationist concerns in the hope that political priorities within the UK would assume a more ecological and less ideological flavour. Wildlife Link soon followed (p. 165) and became an effective means of communication with the politicians. Personal contact between ministers and conservationists was made through the All Party Parliamentary Conservation Committee, administered by the RSNC. Then there were the conservation-minded politicians and the politically motivated conservationists. Neither of these has proved convincing. The Socialist Environment and Resources Association (SERA) was set up in 1973, followed by the Ecology Groups of the Liberals and the Conservatives in 1977.

> As yet, none of the three act as effective channels of access for the environmental lobby . . . . Rather than promoting environmental policies *per se*, each is committed to reconciling environmental issues to the party's philosophy and encouraging environmentally aware people to support that party.
>
> (Lowe and Goyder 1983)

All three of these groups arose in response to the launch of the People Party in 1973, a manifestation of the conservationists' entering the world of politics. Its strategy was based upon the 'Blueprint for Survival', published originally in *The Ecologist* magazine and subsequently in book form. It aimed for zero-growth, decentralisation and self-sufficiency. Such ideologies proved unrealistic, and the party, while changing its name to the Ecology Party and then again to the Green Party, also changed its manifesto. It remained highly idealistic, nevertheless, with aims towards unilateral nuclear disarmament, personal freedom and privacy and an Animal Rights Charter. Its efforts on behalf of nature conservation sometimes seem the most realistic: statutory protection of SSSIs, the establishment of an Environmental Protection Agency and calls for sustainable agriculture. Its finest hour to date came in 1989, when it took third place in the elections to the European Parliament and attracted much media attention for a spell. The main parties wrote off the result as a protest vote, however, and apparently gave it little credence.

The truth is that it was a protest vote. Most people entrust their politics to one party throughout their lives, regardless of policy changes within that party. Indeed, it is very often regardless of any manifesto, but purely for family or traditional reasons! In a general election, these people would not risk 'wasting' their vote on a party they expected to have no hope of success.

The established parties, however, would do well to take note of the protest vote. It is not a negative phenomenon to be taken lightly. In a harmless but meaningful way, those voters expressed their concern for the environment through the ballot box. The parties could only be sure of traditional support continuing if they now embraced environmental issues quickly and convincingly. To a growing number of voters, these issues were the biggest challenge of the times. Many had grown up with economic inflation; never having known anything different, they had learnt to live with it as one of life's realities. Defence seemed less pressing as the shadow of the Second World War receded and political changes in Eastern Europe were easing anxieties in that direction. The environment was on its way to hitting the top of the political agenda.

Only by default, then, might the Green Party expect to lead the country, although it is important that it remains in the political field. It highlights, for the other parties, the policies needed to ensure the future health of the planet. It reflects the growing concern of the electorate. But it is at a local level that the party can be assured of some real success. For, at local level, the environment is a more immediate, more personal preserve. The broader concerns of national politics need not enter here. There may be a need for more housing, larger airports, new industrial sites, but 'in my backyard', the pleasant aspect of my surroundings is of overriding importance. The inherently selfish nature of conservation thus manifests itself most strongly in local politics. It is here that the British Green Party has its strongest influence on the political scene.

On a national level, the party has a lot of work to do before it can hope to win, even by default. It is a young party without an image. As the fear of global destruction spreads, the party may lose its 'cranky' label. Equally, people are learning to live with the fear; the urgency of 1989 has already receded in many people's minds. The British now want to know where the Green Party stands on other issues. The average voter, while meekly accepting that its environmental policies will probably be for the good, has no idea where it stands on education or health. What will its economic policies mean in reality, when stripped of all the conservationist ethics upon which they are based?

So, as the 1990s opened, conservation was nothing more than a minor irritation in the world of politics. Lip-service and empty brochures would suffice to satisfy most people of government good intentions, while an Act of Parliament that pretended a positive attitude to conservation might be used to stifle the government's own advisors.

The empty brochure was the much-vaunted White Paper on the Environment, published in September 1990. Grandly entitled *The Common Inheritance: Britain's Environmental Strategy*, it dwelt more on consideration than on action in approaching environmental problems and made much of those matters that were already in hand. Some 350 proposals looked at tackling vehicle emissions, noise and other nuisances, at giving the Inspectorate of Pollution a degree of independence and at each government department having its own minister responsible for environmental policy. It hoped that new planning controls for the countryside could be brought in to cover farm buildings and to protect hedgerows, and it sought to encourage more energy output from renewable sources. Lord Melchett of Greenpeace described it as 'a woolly wish-list' of possibilities and David Gee, for Friends of the Earth, as 'a discussion programme, not an action programme'.

The voluntary movement had been discussing these issues for years. It was not filled with optimism by a promise to return carbon dioxide emissions to 1990 levels by 2005 and to stabilise them there. It was not convinced of government commitment to the environment by plans to clean up the newly privatised water companies when controls on the still-public electricity industry were put on hold. The statement ignored the really controversial issues and promised nothing in the way of meaningful action. Road pricing and tolls to reduce private motoring mileage were ignored, and there was no response to calls for greater commitment to the railways. Most commentators, of whatever political or environmental inclination, saw it as another empty container, a vehicle to carry a political message that was going nowhere. Bryan Gould, then Shadow Environment Secretary, thought it 'short on policy but long on waffle' and Fiona Reynolds of CPRE 'lamentably lacking in vision, commitment and decision'. Subsequent progress reports have been no more positive, and the same can be said of the equally unimaginative platitudes contained in the environment White Papers of the other main parties.

THE NEW COUNTRY AGENCIES

The subversive Act of Parliament was the euphemistically named Environmental Protection Act, which eventually became law on 5 November 1990 after eleven months of wrangling over detail. Most of it dabbled in typically non-contentious matters such as pollution control and waste disposal, nuisances of noise, litter and stubble-burning, the evaluation of industrial chemicals and the control of genetically modified organisms. These clauses were apparent diversions from Part VII of the Act which decentralised the NCC into separate entities for England, Scotland and Wales, with the amalgamation of NCC and CC duties for the Scottish and Welsh bodies. Although there was much to be said for bringing conservation and amenity interests together after all these years (p. 213), it went totally against contemporary thinking to dismember the official nature conservation unit when environmental matters were increasingly moving onto a continental and global scale.

The English part of NCC became English Nature (EN) in April 1991, remaining a separate entity from the CC. At the same time, the Countryside Council for Wales (CCW) was formed to oversee both nature conservation and the amenity duties previously held in Wales by the CC. The Scottish rump of NCC was left to totter on alone until it merged a year later with the CCS to become Scottish Natural Heritage (SNH).

There the politicians would have left it, but for the uproar within the movement at the cynical weakening of the conservation base on both the national and international scale. The movement was trying to move out on a universal front and here was central government chopping the official body into bits. The politicians demurred and a fourth piece of the new jigsaw was added with the formation of the Joint Nature Conservation Committee (JNCC). It assumed full powers on 1 April 1991, together with its duties to:

- provide advice for government ministers on the development and implementation of policies for or affecting nature conservation for Great Britain as a whole or nature conservation outside Great Britain;

- provide advice and disseminate knowledge to any person about nature conservation for Great Britain as a whole or nature conservation outside Great Britain;
- establish common standards throughout Great Britain for the monitoring of nature conservation and for research into nature conservation and the analysis of the resulting data;
- commission or support research in support of the Committee's responsibilities;
- provide advice to the country nature conservation councils about nature conservation for Great Britain as a whole or nature conservation outside Great Britain.

Another need for which the movement had to fight hard was an independent chairperson for JNCC. Most of the board's other members are delegates from the country units and from Northern Ireland, which had never been represented on NCC.

(The Environment Service had been created within DoE(NI) in 1990, with duties towards nature conservation, amenity, the built heritage, air and water quality, waste management and environmental health. The Council for Nature Conservation and the Countryside was one of its four advisory bodies. The Environment Service was superseded on 1 April 1996 by the Heritage and Environment Agency.)

The political thinking behind the bizarre segregation of NCC must have been as confused as it is confusing to fathom out. Devolution of power to Scotland and Wales is not generally an option even under a regime that likes to delegate. The only credible conclusion is the one that was drawn at the time by so many people: nature conservation had become too strong, particularly in those regions where it mattered most and where, unfortunately for the politicians, 'development' could do with no extra hindrances. Certainly, government wholly failed to convince the movement – and even its own staff – of any good that might have come out of it. Ratcliffe (1989), soon after retiring as NCC's chief scientist, summed up the sourly pessimistic view of most practitioners: 'this will ... be the end of NCC as an effective force for nature conservation in Great Britain'.

Whatever may come from the upheaval in the long term, it is a fact that the exercise was an expensive one both financially and in terms of lost productivity. Preoccupation with the changes clouded the air from July 1989 until at least the first birthday of the new councils. Morale fell and staff performance suffered; impetus and continuity were lost, both in internal administration and in research and practical work. The money and effort wasted would never be recovered.

For, while it was not the sudden and drastic end that many had foretold, little positive good came out of the meddling. The councils' reports, all written by government employees, mention nothing that was achieved because of the split. The increasingly doubtful principles of modern management were embraced by the new units but could more easily have been introduced to the NCC as it was. SNH (1994) maintains that local relations improved in sensitive areas such as Islay, Orkney and the flow country. If this were due to more subtle working practices, they too could have been achieved by NCC, as long as it felt that the cause was not being compromised. To date, there have been no advantages that might be credited to this chaotic political exercise.

On the other hand, the union of nature conservation with amenity interests was a fine move and long overdue. In Scotland and Wales at least, the movement now had no option but to overcome its traditional

Plate 9.1 Nature conservation and amenity use of the countryside were eventually brought together for mutual benefit. Were government plans to make forestry more lucrative about to undo some of the good work? Sherwood in the National Forest. Photograph by Mike Williams. Courtesy of Countryside Commission

divisions. Once this bridge was built, official and voluntary bodies alike quickly came to appreciate the benefits.

> SNH sees access as being complementary to its statutory responsibilities for conservation and for enhancement of the natural heritage of Scotland, and the responsibility to foster greater public understanding and appreciation of it. Indeed, there are countless opportunities where increasing people's enjoyment can go hand in hand with improving degraded areas, increasing people's awareness and, where necessary, protecting the most fragile areas.
>
> (SNH 1995)

> through management of our reserves and our increasing efforts to interpret them for people in towns and countryside alike, we continue to educate and organise ourselves and thus become a more effective force for conservation in Scotland.
>
> (SWT 1995)

All that remained now was to merge EN with the CC in England to build the most effective countryside body possible on that 'national' scale. A six-month study to consider the merger received the tacit agreement of both organisations to co-operate when it was announced in January 1994. Thus ensued another period of uncertainty and faltering productivity. But the government contracted cold feet, perhaps realising just how effective a body it might create and fearing, maybe, another toothed watchdog in the NCC mould. Plans were shelved in October 1994 and both agencies were entreated to work more closely together. This was symptomatic of government's continuing confusion over the new phenomenon called 'environmentalism'.

WOOD AND WATER

The confusion became apparent time and time again. Within a year of splitting the FC into two (p. 179), the government was planning to sell off the Forest Enterprise wing. This was the thinly disguised purpose of the Forestry Review Group, set up in March 1993 to consider other options of ownership for FC holdings, current and possible incentives for investment, and steps to improve the effectiveness of forestry policy objectives. Again there was uproar at the feared loss of amenity and wildlife value as woodlands fell into private hands and the economic objective was restated. Again the politicians were presumably taken aback by the vociferous reaction. The plans to sell were put on hold in July 1994, access arrangements would be expanded and strengthened and there would be higher grants for new planting. But Forest Enterprise was to be put on a more businesslike footing 'with a stronger connection established between the resources used and outputs achieved in all its commercial, recreational and environmental activities' (Ian Lang, Secretary of State for Scotland and MP with lead responsibility for forestry). To that end, Forest Enterprise was to be self-financing by 1995, and 100,000 ha of its holdings would be sold off by the year 2000. Was government baiting the movement, hoping it would not bite back, or was it still so unaware of the strength of feeling?

The water industry in England and Wales had also come up for grabs. The Water Act of July 1989 set up ten regional private water companies. Their responsibilities were water supply and sewage disposal, one very high in the public mind, the other preferably kept out of sight.

Conservation and amenity interests were unhappy with the potential results. The CPRE, in conjunction with the RSPB and WWF, outlined the fears in the 1988 report: *Liquid Assets*. One otherwise largely ignored aspect was the amount of land that the new companies would take on: 182,000 ha, much of high scenic value. All that to fall into private ownership? So the government published a Code of Practice for Conservation, Access and Recreation, laying a duty on the companies to disclose all land sales to the water watchdog, OFWAT.

The greater concern, however, was the level of commitment they would show to environmental matters over the holdings they retained. With shareholders to satisfy and an initial ten-year investment of £24 billion to make, how much cash and concern could they be expected to put into good land management, for example, or pollution control? As it was, they were inheriting years of inefficiency and bad practice and a dismal record of secrecy and statistical prevarication (p. 144). Again, the government thought it politic to make a concession in the form of a National Rivers Authority (NRA) to oversee the more public-minded, less lucrative aspects of river regulation and management, such as pollution control, flood defence, fisheries and water resources.

Amongst the conservationists, hopes of impressive performance from the NRA were not high, yet it quickly proved its worth. Immediately, it reviewed discharge consents, and many companies were refreshingly forthcoming in reducing discharges when approached. With a growing public stance against pollution, and the options of increasingly clean technologies, it was in their interests to do so. Besides, monetary charging for effluent discharging was introduced in 1992, by which time the NRA had shown its willingness, on occasions at least, to prosecute polluters ... and with a high rate of success.

The water companies, to date, have achieved little more than increased capital investment in the industry. Whatever else we may say about government, it was a smart move to get out of the water when it did. While the public had taken the industry for granted, the evils had been able to pile up unnoticed. But in the 1990s, water made a big splash! Droughts seemed to become an annual problem; water shortages lasted the year round. The Lake District sent water by road to Yorkshire in

Plate 9.2   Concern over impending changes in the water industry extended to
182,000 ha of valuable land which also appeared to be 'up for grabs':
Wastwater and Great Gable in the Lake District National Park. Photograph by
Mike Williams. Courtesy of Countryside Commission

November 1995, and talk was heard of literally shipping it between
European nations by sea.

This new awareness must now be used to encourage a reduction in the
demands we make upon water. That would be all to the good of nature
conservation. More than that, it will fall to the official agency to push the
water companies towards greater environmental awareness. The NRA
apparently had the ability and attitude to do so. It will be interesting to
see if its successor, the Environment Agency, develops the same outlook.

For organisational changes were again being imposed. In July 1991, the
proposal was heard to merge the NRA with HM Inspectorate of Pollution.
HMIP had only been in place for four years and, under the Environmental

Protection Act, had just been tasked to oversee Integrated Pollution Control (IPC) in relation to thirty-three industrial processes which involved the use of specified dangerous substances. It had taken fifteen years to get IPC into operation, and the system relied upon industry's offering all the right information in the correct fashion.

On 1 April 1996, the NRA and HMIP were integrated with the local waste regulation authorities in the Environment Agency for England and Wales. In Scotland, the industry had not been privatised and water services were still within the remit of the local authorities, with river purification boards monitoring pollution. The boards merged with HMIP and the local waste regulators to form the Scottish Environmental Protection Agency (SEPA). Conservationists welcomed the move towards co-ordinated pollution control, but had been disconcerted by the draft proposals in the Environment Bill of October 1994. Although the main duty of the new bodies would presumably be to protect and enhance the environment, and although financial objectives were outlined, nothing was written in to cater for environmental objectives: no requirement to clean up polluted habitats, no suggestion even in vague terms of any duty towards the conservation of nature. The movement did manage to force an amendment on nature conservation but the other omissions remained and, in Scotland, the RSPB viewed it as 'a missed opportunity to provide a more integrated approach to water management' (RSPB 1995).

The Environment Act finally reached the statute books late in 1995. Kinnersley (1994) gives one reason for its inordinately slow passage as 'the view that environmental problems are slipping down the political agenda from the "white heat" they seemed to generate a few years back'. White heat or not, the particular problem of pollution in all its forms was something that the politicians could be ignoring only because of their acute embarrassment. It was worse now than in the bad old days. In England and Wales, the number of reported pollution incidents had risen from 12,600 in 1981 to 25,500 in 1989 and 31,680 in 1992. Some of the increase can be attributed to a keener awareness of the problems, but does that make the figures any more palatable? Major incidents reduced by 40 per cent into the early 1990s but at 388 in 1992 were still happening more than once a day. Twenty-eight per cent of all reports were sewage related, 26 per cent were industrial and 12 per cent were agricultural. Animal waste can be one hundred times, and silage two hundred times, more damaging than human sewage, and MAFF was actively pursuing the introduction of farm waste management plans.

Yet more stretches of river were deteriorating than were improving (Table 9.1). The River Restoration Project was set up in December 1992 to tackle the most degraded lengths. In the summer of 1995, tens of thousands of fish in English Midland rivers died as a result of pollution coupled with thunderstorms and high temperatures which reduced oxygen levels. Schemes to reduce nitrate pollution (p. 247) were stepped up and

Table 9.1 Water quality in the 1980s

Quality change between 1985 and 1990 (England and Wales)

Type of water	Improved (km)	Degraded (km)
Rivers	4,444	5,886
Canals	179	371
Estuaries	18	83

Freshwater rivers and canals (England and Wales)

Class	1980 (km)	(%)	1985 (km)	(%)	1990 (km)	(%)
Good	28,500	69	27,460	67	26,944	63
Fair	8,670	21	9,730	24	10,750	25
Poor	3,260	8	3,560	9	4,022	9
Bad	640	2	650	2	662	2
Total	40,630		41,390		42,378	

Estuaries (England and Wales)

Class	1980 (km)	(%)	1985 (km)	(%)	1990 (km)	(%)
Good	1,870	68	1,860	68	1,805	66
Fair	620	23	650	24	655	24
Poor	140	5	130	5	178	7
Bad	110	4	90	3	84	3
Total	2,730		2,730		2,722	

Source: NRA (1991), 'The Quality of Rivers, Canals and Estuaries in England and Wales', quoted in Kinnersley (1994)

Freshwater rivers and canals (Northern Ireland)

Class	1990 (%)
Good	72
Fair	24
Poor	4
Bad	0

EN stated its intention to notify twenty-five English rivers (3 per cent of the total length) as SSSI. The heat was still on when it came to water.

The quality of drinking water was improving, though it still fell short of EC standards in many areas. The government baulked at the thought of the companies' increasing water rates too sharply. Yet, as Kinnersley (1994) wryly points out, a growing number of consumers were paying one thousand times more for bottled water than tap-water and still had the pleasure of having to carry it home!

Pollution of the oceans went on apace. Britain sent to sea 1,500 million litres of raw or undertreated sewage every day, murkily undermining all efforts to clean up the beaches. EC Blue Flag Awards (p. 184) continued to prove elusive; eighteen were awarded in 1995. The Tidy Britain group introduced a UK-only scheme of Resort Beach Awards and Rural Beach Awards which were less stringent in their requirements. Maybe 1,500 tonnes of oil were legally discharged into the North Sea every year. The illegal total probably approximated 60,000 tonnes. *MV Braer* shed 84,000 tonnes in one go off the Shetland coast in January 1993. High winds whipped up the sea and swept most of it under the surface so that neither the worst fears of the conservationists nor the greatest embarrassment in official circles was realised. The case of the *Sea Empress* was somewhat different. Holed whilst entering the well-mapped mouth of Milford Haven in February 1996, her cargo of oil was allowed to spread unsightly death and destruction across much of the Bristol Channel. Maybe the British are pragmatic enough to accept that such accidents will happen, but do they know that ten times as much oil is routinely, if unlawfully, discharged during tank-cleansing operations than by accident?

It all added up to an horrendous catalogue of 'materials in the wrong place' and, for the politicians, public concern was getting serious. More people were now more worried about pollution than about any other environmental concern, new or old. It had outstripped nuclear waste disposal as a source of disquiet and was deemed to be more pressing even than the fraying of the ozone layer and global warming. Acid rain lagged far down the league table. The ratings were obviously tempered by the feeling that, while global and atmospheric problems seemed insurmountable, it was patently possible to do something about pollution.

And so, it was all the more telling that overfishing ousted pollution as the main concern of the fourth North Sea Conference in June 1995. The economics of industry remained the priority of the politicians. Yet this industry, more than many, relied upon a clean environment in sustainable use, and it was vital that the participating nations take a serious view of depleting stocks. After several years of plummeting seabird populations, attributed to lack of marine food species, the RSPB had succeeded in getting onto the statute books the Sea Fisheries (Wildlife Conservation) Act 1992, which stipulated that fishing interests should take account of marine wildlife and its conservation. In September 1993, an inter-agency fisheries task force had been launched. But a wreck of 75,000 starved seabirds, washed onto North Sea coasts in February 1994, only served to remind delegates of the severity of the problem. Yet the conference came up with nothing in the way of positive action. It decided to consider sustainable fishing and habitat protection at a later meeting in 1996. Incidentally, it would also take some steps to counter oil pollution.

URBAN AND INDUSTRIAL VENTURES

While the water companies were coming to terms with their inheritance of polluting systems, and farmers were learning to live with waste management plans, industry was continuing its efforts to clean up its image. Company policies made much of pollution control and cleaner technology, reduction of waste and increased recycling. Positive improvements were seen. The electricity companies began undergrounding low-voltage cables, in spite of the much higher costs involved and their traditional reluctance to do so. After being fined £1 million for a bad oil leak into the River Mersey in 1990, Shell invested £30 million in new processes to cut pollution.

The Environment Council launched its Business and Environment Programme in 1989, and the heavy industrial companies on Teesside set up an Industry and Nature Conservation Association (INCA). The NCC subsequently developed the INCA concept in other areas as a way of bringing the two interests together for mutual benefit and understanding before entrenched positions could be taken up. The International Chamber of Commerce established a World Industry Council for the Environment in 1993 to push green issues to the highest levels in industry worldwide. The EC's Eco-Management and Audit Scheme encouraged the highest environmental standards in industry. Five British firms, in 1995, were the first to sign up. A group of British companies formed a Corporate Forum for National Parks. The conservation movement was pleased to encourage the trend. The Environmental Protection Branch of the CC, set up in 1990 in order to broaden the sphere of the commission's work, was seen as a way of forging new links with 'industry's increasing interest in green issues'. For, while much industrial 'greening' had previously been for cosmetic or marketing reasons, there was nonetheless considerable potential in developing a strain of environmentalism within the industrial world.

Meanwhile, traditional sponsorship remained a high profile public relations strategy. The 1994–5 report of the CC alone listed ninety-nine companies that had assisted during the year towards the provision of almost £2 million. While actual amounts were small in terms of company turnovers, they often represented, out of all proportion to the pounds and pence involved, the power of small groups and individuals to realise personal dreams close to their hearts and vital to their local surroundings.

For small-scale, localised achievements are one of the most important components of the current conservation and amenity movement. From the 1970s onwards, central government increasingly delegated involvement in a wide range of issues to the more local scale. Personal ownership of an initiative leads to better targeting of specific problems – those identified by the local community – and keener determination to achieve positive results. Conservation has gained more than many other issues from this appeal to the 'nimby' attitude. 'Not in my back yard' and its positive counterpart, 'yes, in my back yard', ignites an enthusiasm for the good things in our local surroundings and impels a sometimes astonishing resolve to retain or achieve them. It is, if you like, a form of the selfishness upon which the movement is founded. The mode of government that devolves responsibility, together with the range of schemes that now offer funds and advice – many of them financed by industry – has empowered groups and individuals to take positive action for their local environment. Greater leisure time, some of it enforced by unemployment, has added to this tendency to volunteer, to participate for reasons other than financial reward (p. 255).

The result has been the creation or rehabilitation of some wonderful havens for wildlife and recreation. The whole achievement is proving greater than the sum of its parts. Most of these pockets are integral to a community; a breathing space where it is most vital, where the spiritual need is greatest and the capacity to convert others to the cause most promising. But this thickening network of green spaces is valuable, too, as a series of stepping stones or developing corridors between other designated wildlife sites.

Thus the Shell Better Britain Campaign (p. 187), relaunched in 1995 after a ground-breaking first quarter-century, concentrates on localised activities that can help towards global sustainability. So pandemic is its view of the environment as a single entity that grants may be given to neighbourhood watch and even health schemes as well as to more traditional conservation ventures. Its annual *Interactive* has filled the gap left by the *Conservation Annuals* (p. 188) as a guide to some of the impressive range of local enterprises that are being pursued all over the country.

King and Clifford (1985) give other examples, supported by good practical advice on doing it yourself. They founded Common Ground in 1983 to encourage an appreciation of important features of local landscapes through ventures such as the 'Parish Maps Project' and, more recently, 'Local Distinctiveness'. By distilling the essence of the local landscape, appeal is made precisely to the personal commitment that will stand up and fight on behalf of it. By preserving local character, we retain a healthy diversity on the national scale, hopefully avoiding the charge that all towns or all villages look the same. Common Ground's other fascinating and original initiatives have included a 'Trees, Woods and the Green Man' project and the 'Save Our Orchards' campaign.

As well as industry, the official and voluntary conservation bodies have realised the rewards that might be reaped by enabling others to satisfy their environmental hopes. EN piloted 'Community Action for Wildlife' in 1990 which, in its fourth year, funded 170 small-scale ventures. The Prince's Environmental Trust for Wales aims 'to encourage and assist practical environmental action and understanding by communities throughout Wales'. Rural Action for the Environment offers finance, training and advice to local projects. Its national co-ordinating group consists of all the official wildlife and amenity bodies, together with the RSNC, the BTCV, the Shell Better Britain Campaign, Action with Communities in Rural England and the National Council for Voluntary Organisations. By June 1995, it had chalked up 29,000 voluntary work-days on 1,250 projects.

The BTCV administered the Barclays UrbanLife Campaign. The CPRE launched an Urban Footprints initiative. Scottish Countryside Around Towns catered for eighteen projects in 1994. The following year, the Scottish Wildlife Trust took on the management of Cumbernauld's 320 ha of green space. There was an Urban Wildlife Partnership north of the border, and another in England and Wales that consisted of about one hundred groups, wildlife trusts and local authorities, brought together in 1992 under the RSNC umbrella. The Urban Regeneration Agency was formed to revitalise derelict land by using the complementary resources of local authorities, voluntary bodies and private companies.

The Groundwork Foundation was co-ordinating the work of the Groundwork Trusts as they restored landscape and wildlife habitat on derelict and waste land in towns and cities throughout Britain. The first trust had been established in December 1981 at St Helens and Knowsley. It found so much good work to do that similar groups soon sprung up in other towns and cities. By 1994, there were thirty-four trusts in England and Wales alone, and the foundation was put in place in 1985. Local beginnings really could lead to national movements.

Understandably, many of these ventures have targeted urban problems, where communities are largest and the need for action more evident. The battle to mend old scars remains as necessary as ever. After the 1970s saw an increase in urban blight, there was an 11 per cent reduction in England during the 1980s. Three thousand hectares were being reclaimed annually. Yet the war was not being won. The figures for derelict land in England, Wales and Scotland – 68,000 ha in 1988, 70,000 ha in 1995 – equated with those for 1969 (p. 159). Attitudes towards urban land use are still too lax, and the need for urgent remedial work looks set to continue well into the new century.

A NEW NATIONAL FOREST, A NEW NATIONAL PARK

While so much was going on in the towns and cities, other exciting developments were happening on their doorsteps. Forestry was taking its place as a viable use for derelict or even agricultural land on the urban fringes. The acorn of the idea for a new National Forest was sown by the CC in its December 1987 report, *Forestry in the Countryside*. The site was chosen from a short list of five in 1990: 502 sq km, the size of the Isle of Wight, based on the ancient forests of Needwood and Charnwood and bounded by the major conurbations of the West Midlands, Leicester and Derby/Nottingham. Much of the region was bland agricultural land; much was blighted by industrial dereliction in the form of worn-out gravel workings and the sad relics of the British coal industry. In all respects, human and natural, it was an area that cried out for rejuvenation. With tree cover set to reach 50 per cent in the ensuing twenty five years, the forest was seen

Plate 9.3 The National Forest would reclaim large areas of industrial dereliction and bland agriculture and turn it into a valuable recreational resource: the former coal mine at New Clipstone. Photograph by Mike Williams. Courtesy of Countryside Commission

as a major recreational and environmental resource for the future. After four years of intense planning and development by the CC, FC and local authorities, during which time the first half-million trees were planted, a public company was formed to carry the work forward.

Additionally, twelve community forests were announced. These too would be multi-purpose, urban-peripheral sites (Fig. 9.1) where landscape and wildlife, amenity and employment would all be given a better chance. At the outset, tree cover averaged less than 7 per cent, which was itself below the national average, but it was set to increase to 30 per cent with community forest status. Totalling 4,660 sq km, community forests approximate half the area of National Parks in England.

Similarly, SNH initiated the Central Scotland Woodlands Scheme to promote major multi-purpose forestry on degraded land in the urbanised Central Belt. Once established, the venture was given its independence from SNH. 'The Millennium Forest for Scotland' is a consortium of conservation interests which hopes to double the country's area of native woodland by 2000. A major element of the project is to regenerate the Caledonian Forest, which has been reduced by a quarter since 1960 alone.

The CC and CCW were also busy in the National Parks where, for decades, concern had been growing over the lack of real protection from development, quarrying and military use. The 1974 Sandford report of the National Parks Policies Review Committee (p. 157) had called upon the government to extend to all the National Parks the provisions of the 1950 Landscape Areas Special Development Order (LASDO). In this way, some control could be held over agricultural building development in these areas. Sandford also recommended planning controls upon forestry and a power of compulsory purchase of land in the National Parks with a particular value to conservation. The latter two proposals were turned down flat. On agricultural development, political promises were made but not fulfilled. A backward step followed in 1981 when legislation was introduced that enabled park authorities to award grants to landowners adversely affected by LASDO. The opportunity to at least appease the amenity movement by extending LASDO to all National Parks was allowed to slip by. The farmers were clearly in control.

The CC maintained the pressure and, in autumn 1985, launched its two-year awareness campaign, 'Watch over the National Parks'. Finally, in 1986, twelve years after Sandford had reported, LASDO was extended to all National Parks and made to apply to new roads as well as buildings, to forestry as well as farming. Planning authorities were given twenty-eight days in which to call for a full application if they wished. Amenity interests responded by calling for an extension to all upland regions and all AONBs, but they were to be disappointed.

In 1987, again in response to apparent public concern, government mooted the idea of Landscape Conservation Orders (LCOs). This measure would have provided park authorities with a last-ditch power to prevent

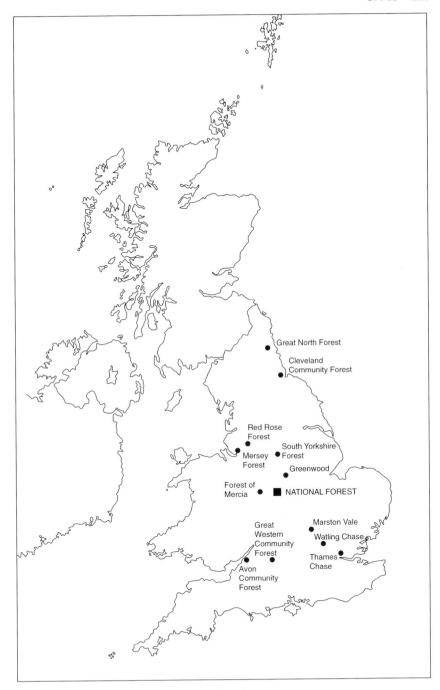

Figure 9.1 The National and Community Forests

the destruction of natural beauty where all other negotiation had failed. But now the pressure faltered and no political action followed the words.

Public interest in the National Parks had, however, been stirred and, in March 1988, the Norfolk and Suffolk Broads were given a new Broads Authority which was to run the region on National Park lines for recreation, navigation and wildlife. Although the Broads did not technically become a National Park, it was the first new area since 1957 to effectively receive the same treatment and it reopened the debate on other potential regions, such as the New Forest, South Downs, North Pennines and Cambrian Mountains.

The National Parks Review Panel convened in 1990 under Professor Ron Edwards. Its report in the following year, *Fit for the Future*, carried 170 recommendations, the most important of which were that all parks should be run by independent authorities under a central co-ordinating body, and that a statutory 'Silkin test' should apply to all major development plans within the parks and, only if there were an overriding and unavoidable national need for it, should a development go ahead. Edwards called for a duty within the parks towards wildlife and cultural heritage, and for all ministers, government departments and public agencies to further the purposes of the parks in all the activities they undertook. It also continued the calls for the New Forest to be recognised as a National Park or equivalent.

The CNP had hoped that the further areas mentioned above would have been highlighted. It was disappointed that greater emphasis was not given to the need for effective park funding and for preventing tourism from developing along lines which were inimical or insensitive to park purposes.

Official response came early in 1992. It promised independent authorities, and special protection for the New Forest. But the enabling Environment Bill was making slow headway through the soup of government inertia, and it began to seem that county councils were bound to lose interest in their National Parks. After all, their involvement in them was soon to end, if indeed the very councils themselves did not disappear in the threatened reorganisation of local government. Fearing a spell of utter vulnerability for the parks, Lord Norrie, Vice-President of the CNP, promoted a Private Member's Bill that would at least have put the free-standing authorities in place. It was blocked in the Commons in May 1994.

So Part III of the Environment Act 1995 was the first legislation to address the running of the National Parks since their inception in 1949. An independent authority would run each of the ten parks, with more local representation on the English boards but, strangely, with little change in the make-up of the Welsh authorities. This disquieted those who already felt that the three Welsh parks had suffered from falling resources since their segregation in 1991; certainly there was no apparent reason for treating them differently unless it was to weaken them. An Association of National Parks (ANP) was established as the central forum for the individual authorities, while the CNP, of course, continued to represent the voluntary environmental and amenity organisations.

Under the Environment Act 1995, the revised purposes of National Parks were:

- to conserve and enhance the natural beauty, wildlife and cultural heritage of the areas;
- to promote opportunities for the understanding and enjoyment of the special qualities of those areas by the public.

In the event of conflict, the first purpose was to take precedence. National Park authorities were given a duty to foster the social and economic well-being of local communities. This economic duty caused grave concern to many who saw it as a requirement, or at least an excuse, to actively promote development that might not have the best interests of the parks at heart. Public authorities were to have regard in their work for the purposes of the National Parks, but ministers and government departments were spared the same duty. There were still no strict planning restrictions for the parks; the military, road-building, quarrying and all the rest were still too close to the heart of government. Moreover, although the promise had been repeated as recently as July 1994 that the New Forest would receive the same planning protection as the National Parks – whatever that was – there was nothing in the Act to bring it into effect. While reaffirming the existence, perhaps even the desirability, of National Parks in England and Wales, the politicians had managed once more to sidestep the main issues.

North of the border, renewed interest in National Parks manifested itself in the 1990 CCS report, *The Mountain Areas of Scotland*. This proposed park status for six areas: Assynt-Coigach, Wester Ross, Cuillin Hills, Ben Nevis and Glencoe, Loch Lomond and Cairngorm. This last would cover nearly 400,000 ha and have much in common with continental parks. But official will was lacking again. Natural Heritage Areas were proposed as a sop, promoted as a 'multi-purpose designation covering land-use and management, landscape and nature conservation, amenity, access and enjoyment'. So it was no surprise that the Scottish Council for National Parks was reconstituted as an independent voluntary body in 1991. In that year, the Natural Heritage (Scotland) Act legislated for the new areas but, in 1995, SNH had yet to present to government for approval, its criteria for their selection and designation. It was already clear that, even though large areas might eventually come under the tag, there would be no real protection as a result.

The 1949 Act had catered for the designation of AONBs as well as National Parks. By 1990, thirty-eight had been recognised in England and Wales, covering 20,000 sq km. Nine in Northern Ireland extended to almost 2,000 sq km. Certainly they helped to fill the gaps in regions with

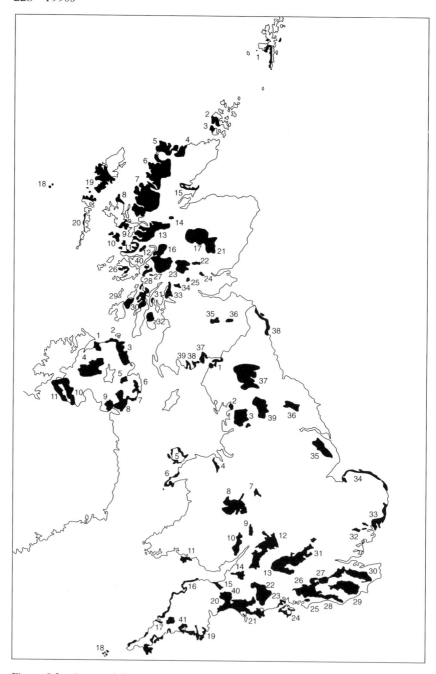

Figure 9.2   Areas of Outstanding Natural Beauty (England, Wales and Northern Ireland) and National Scenic Areas (Scotland) 1996
National Scenic Areas, Scotland: 1 Shetland; 2 Orkney, West Mainland; 3 Hoy; 4 Kyle of Tongue; 5 North-West Sutherland; 6 Assynt-Coigach;* 7 Wester Ross;*

no National Parks but they had always been the poor cousins of the parks. The CC did not have the resources to promote them and, of all the questionable designations, AONB stagnated more than most. Now, however, prompted by public opinion, local authorities began to take an interest in their AONBs. The CC was quick to update its approach and published *Areas of Outstanding Natural Beauty: A policy statement* in 1991. It called for more realistic funding and for the appointment of AONB officers and Joint Advisory Committees to manage and protect these areas with more local commitment. By the end of 1995, when AONBs in England and Wales totalled forty-one and exceeded 21,000 sq km, twenty-six advisory committees and twelve AONB officers were in post.

Inevitably, various forms of administration were tested. In the North Pennines AONB, a Tourism Partnership was formed between the CC, Rural Development Commission, tourist boards, local authorities and the private and voluntary sectors. It aimed to promote the economy of the region in a way that was sensitive both to local social needs and to the environment. More innovative was the establishment on 1 April 1992 of the Sussex Downs Conservation Board, a statutory committee of the thirteen local authorities and the CC which, one year later, assumed full powers and duties to protect, conserve and enhance the environment,

8 Trotternish; 9 The Cuillin Hills;* 10 The Small Isles; 11 Morar, Moidart and Ardnamurchan; 12 Loch Shiel; 13 Knoydart, Kintail and Glen Affric; 14 Glen Strathfarrar;* 15 Dornoch Firth; 16 Ben Nevis and Glencoe;* 17 The Cairngorm Mountains;* 18 St Kilda; 19 South Lewis, Harris and North Uist; 20 South Uist machair; 21 Deeside and Lochnagar; 22 Loch Tummell; 23 Loch Rannoch and Glen Lyon; 24 River Tay (Dunkeld); 25 River Earn (Comrie to St Fillans); 26 Loch na Keal, Isle of Mull; 27 Lynn of Lorn; 28 Scarba, Lunga and the Garvellachs; 29 Jura; 30 Knapdale; 31 Kyles of Bute; 32 North Arran; 33 Loch Lomond;* 34 The Trossachs; 35 Upper Tweeddale; 36 Eildon and Leaderfoot; 37 Nith Estuary; 38 East Stewartry Coast; 39 Fleet Valley; 40 Loch Linnhe. Areas assigned asterisks above are recommended National Parks by CCS (1990)
Areas of Outstanding Natural Beauty, England and Wales: 1 Solway Coast; 2 Arnside and Silverdale; 3 Forest of Bowland; 4 Clwydian Range; 5 Anglesey; 6 Lleyn; 7 Cannock Chase; 8 Shropshire Hills; 9 Malvern Hills; 10 Wye Valley; 11 Gower; 12 Cotswolds; 13 North Wessex Downs; 14 Mendip Hills; 15 Quantock Hills; 16 North Devon; 17 Cornwall; 18 Isles of Scilly; 19 South Devon; 20 East Devon; 21 Dorset; 22 Cranborne Chase and West Wiltshire Downs; 23 South Hampshire Coast; 24 Isle of Wight; 25 Chichester Harbour; 26 East Hampshire; 27 Surrey Hills; 28 Sussex Downs; 29 High Weald; 30 Kent Downs; 31 Chilterns; 32 Dedham Vale; 33 Suffolk Coast and Heaths; 34 Norfolk Coast; 35 Lincolnshire Wolds; 36 Howardian Hills; 37 North Pennines; 38 Northumberland Coast; 39 Nidderdale; 40 Blackdown Hills; 41 Tamar Valley
Areas of Outstanding Natural Beauty, Northern Ireland: 1 North Derry; 2 Causeway Coast; 3 Antrim Coast and Glens; 4 Sperrin; 5 Lagan Valley; 6 Strangford Lough; 7 Lecale Coast; 8 Mourne; 9 Ring of Gullion; 10 Erne Lakeland (proposed); 11 Fermanagh Caveland (proposed)

Plate 9.4   For forty years, the underfunded AONBs had languished under a meaningless designation, vulnerable to all the pressures in the countryside: quarrying in the South Downs AONB. Photograph by Matthew Stevens. Courtesy of Countryside Commission

promote quiet public enjoyment, and foster sustainable economic and social development in the area. It employs a planning officer and has a statutory right to be heard in planning matters, while remaining separate from the local planning process. The CC nominates a third of its members, finances half its costs, and sees it as a model for the future management of all AONBs.

THE WIDER COUNTRYSIDE: PLANNING AND ACCESS

The busy CC was also working hard to ensure a viable future for the wider, undesignated countryside. Its report of October 1991, *Caring for the countryside: A policy agenda for England in the Nineties*, contained almost one hundred proposals for an integrated strategy that would ensure 'a sustainable, multi-purpose countryside – beautiful, environmentally healthy, diverse, accessible and thriving'. It called for a co-ordinated national policy on rural land use, towards which the government took a first hesitant step in November 1994 when it began the consultation procedure for a Rural White Paper upon the practical effects of sustainable use of the countryside on the rural economy, society and environment into the twenty-first century.

To assist the process of clarifying just what it was in the countryside that people wanted to preserve, the CC launched the New Map of England in 1992, to pilot different methods of landscape assessment. The scheme subsequently became part of the Countryside Character Programme, together with EN's Natural Areas concept and the work of English Heritage on buildings and monuments. Like the innovative work of Common Ground (p. 221), these projects sought to distil the essence, the fundamental characteristics of the environment in order to pinpoint, region by region, those aspects that featured in the consensus view of the ideal countryside. The final result would be a picture that all countryside interests should wish to protect. But it might, of course, tend towards the lowest common denominator.

> Natural Areas reflect the geological foundation and natural systems and processes operating in the different parts of England. They are extensive and have a unity and sense of place to which people can relate; in them land and water management can be integrated . . . . Within such areas we propose that agreed levels of nature conservation interest will be secured – and that local communities will be intimately involved in defining the goals and achieving them.
>
> Natural Areas provide a logical framework for existing recognised areas . . . . Based on this framework, we believe a new approach to nature conservation can emerge.
>
> (EN 1993)

The White Paper on *Rural England: A Nation Committed to a Living Countryside* was launched on 17 October 1995 and gave government pledges to conserve natural resources, reverse the decline of wildlife, increase opportunities for recreation, and respect the diversity of landscapes and the character of rural towns and villages. It also stressed the need for communities to remain economically viable. Its principles for future countryside policies were based upon:

- sustainable development to meet current needs without compromising the ability of future generations to meet theirs;
- open dialogue between communities and all levels of government to reconcile competing priorities;
- respect for local landscape and wildlife distinctiveness and the economic and social diversity of the countryside;
- improved understanding of economic, environmental and social trends.

The CC was entreated to work more closely with the Rural Development Commission in order to reconcile the often divergent goals of economic and environmental improvement. Economic development, it seemed, would always be detrimental to the environment. In 1995, a policy statement from the Landscape Institute, sponsored by the CC, SNH and the CCW, called for landscape and visual impact assessment on all significant projects. There had to be a way to bring these apparently competing causes together.

Much of the concern, of course, arose out of continuing reservations over the planning machinery. For all the structure plans and development plans, it still appeared a piecemeal process, totally *ad hoc* and dependent on the shifting sands of personalities and persuasions.

> It is characteristic of the majority of such [planning] enquiries – window dressing or not – that the inspector is an expert on road construction, camber, traffic flow, tarmac and topsoil costs and so forth, but has neither knowledge nor interest in aesthetic considerations, natural beauty, wildlife and disappearing species. It is quite extraordinary that, if an area is scheduled as one of outstanding natural beauty, we should not have had two inspectors – one appointed by the Department of Transport, the other by the office of the Minister for the Environment or Science – who would have submitted a joint report when considering this proposal: a new road gouged across a scheduled area of chalk downland like the Tring bypass. At least the dual interests could then have been considered together.
>
> (Rothschild 1992)

How could that be in these days of sustainability and environmental interdependence? Everybody else was talking in such terms; why not the planners? Surely the sciences of design and engineering were sufficiently advanced to be able to harmonise roads and buildings with landscapes in such a way as to actively enhance local distinctiveness and national diversity?

If the science was strong, the will was weak. Planning guidance was reviewed from time to time, but the actual mechanisms stayed unaltered. Crown bodies and government departments, which controlled 10 per cent of SSSI area in England alone, remained exempt from planning restrictions. There were even moves to relax controls within the 1.65 million ha of Green Belt. The CPRE and others successfully resisted the tendency and government policy was restated in 1994, but there seemed to be an increasing willingness on the part of local authorities to bend the rules. Hertfordshire County Council wanted to trade off its participation in a Community Forest for a relaxation of the Metropolitan Green Belt. In Cheshire, a three-year battle against the threat of development on 3,200 ha of Green Belt around Chester was won in January 1992, but the war continued in other areas.

Official attitudes lagged far behind the conservationist ethic. When plans to build the bridge to the Isle of Skye were found to be jeopardising the holts of thirty otters, the Scottish Office was reminded of its duties under the Bonn Convention and the Wildlife and Countryside Act. It replied that, quite apart from its being exempt from the Act, the damage would not be illegal because it was the incidental result of a lawful operation. Presumably that was its genuine considered opinion. Looked at in that light, planning approval could override almost every Act and convention. The planning machinery was in desperate need of urgent and intense review.

So too was the continuing homage shown by government to the god of economics and development. The Channel Tunnel caused unprecedented environmental damage ... and to what end? By 1995, it was losing £2 million a day. How could all the authorities involved have been taken in by such monumental miscalculation? Only by a kind of unquestioning, reverential blindness.

The CC and CPRE and their sister organisations in Scotland and Wales were the main protagonists in the fight to tighten up planning. One or two minor concessions were achieved, more of a theoretical than practical nature. Under the Planning and Compensation Act 1991, the traditional assumption in favour of development was finally overturned – on paper at least. Planning Policy Guidance stated that 'the sum total of decisions in the planning field ... should not deny future generations the best of today's environment'. Sustainability had apparently found a way through the sacred portals of planning. The Act was the first piece of primary legislation specifically to mention Environmental Assessments (p. 195). It also reiterated the overall authority of county and district development plans. Most agricultural buildings were at last to be notified to planning authorities and the DoE began to give serious thought to ending Crown exemption. Tentative thoughts, barely uttered, even suggested that future developments should minimise the need for day-to-day travelling on routine business. What was needed now was a fundamental change of attitude amongst the planners, so that planning in the future might at least become planning for the future.

Public access to the countryside continued to take up much time in the offices of the CC, CCW and SNH. In England and Wales, there are 225,000 km of footpaths and 33,000 km of bridleway; there are byways and commons, country parks and pocket parks, forest parks and forest trails, and an increasing amount of land being opened up for public use. The CC view was that 'it is completely unreasonable in the last decade of the twentieth century for a small number of landowners to expect to be able to bar the public from their land' (Blunden and Curry 1990) and it felt that correct behaviour from the public would only come after access was granted. In 1987, it published its aim to have all rights of way legally defined, well maintained and publicised by the end of the century. In 1993,

Plate 9.5   Planning and access remained two priorities for the official amenity bodies. Landowners were encouraged to renovate suitable buildings and allow their use for simple overnight shelter: camping barn at Catholes, Sedbergh in the Yorkshire Dales National Park. Photograph by Ian Carstairs. Courtesy of Countryside Commission

it reminded local authorities of the goal by introducing the 'Milestones' initiative by which they could gauge their progress. The CCW instigated a survey of footpath condition. The 1990 Rights of Way Act tightened control over the restoration of paths after ploughing, and the CC's Parish Paths Partnership, launched in 1992, encouraged the active upkeep of footpaths at local level. By 1995, over a thousand parishes had become involved.

Meanwhile, new routes and trails were being blazed. The CC and CCW were active in promoting the National Trails which, by 1995, had risen to eleven in number and 3,328 km in length. Plans were well advanced for further additions: the Thames Path, the Hadrian's Wall Trail and the upgrading of the Cotswold Way to National Trail status.

A newer innovation was the networking of routes, rather than the traditionally linear form. Salford City Council and North Tyneside Metropolitan Borough Council were amongst those to renovate disused railway land for commuting and recreational networks. The London Walking Forum aimed to develop inner and outer orbital routes and a web of paths within the capital which would exceed 1,000 km in total. Five interlocking trails in Cannock Chase, designed particularly with the disabled in mind, were evidence of the continuing high priority given to a countryside for all. The CC was funding an 'Access For All' advisory service, with the Fieldfare Trust as its consultants. The Scottish Countryside Rangers Association organised a 1992 conference on 'The Countryside For All', following on from which SNH ran a campaign of the same name.

In Scotland, however, the whole question of access was very different. Few paths were official; a time-tested regime based on tacit agreement and mutual respect had allowed access across vast areas of open land to nobody's detriment. Perhaps there were too many people south of the border to allow such a system to work there ... or perhaps we were by then too burdened with legal niceties to see the potential? But the fear in Scotland was that landowners might seek to move further towards the southern position as pressures in the countryside grew. This would jeopardise the traditional arrangements while there were no legal provisions to make up the shortfall. In fact, SNH felt that opportunities for access, especially on lower land, 'in some respects ... compare unfavourably with those available in Europe and indeed in other parts of Britain' (SNH 1995). The Countryside (Scotland) Act 1967 had legislated for long-distance footpaths, but not until 1980 had the West Highland Way been designated as the first. By 1995, only three more had been added – the Southern Upland Way, Speyside Way and Great Glen Way – giving a total length of 500 km.

The SNH review of access concluded:

- there is still insufficient recognition of the importance of open-air recreation to people in a modern society;
- the level of planning for access, management of access and investment in access is too low;
- the countryside is often neither very accessible nor welcoming, so that people can find it difficult to know where they are free to go;
- growth in recreation in the countryside affects land management operations and there are worrying indications of increasing friction;
- people are uncertain about whether they have any legal rights to be on privately-owned land;
- the arrangements in the Countryside (Scotland) Act 1967 for providing for formal access through footpath agreements have been little used, and new approaches are required.

Through its Paths for All initiative and an Access Forum, SNH hoped to overcome these concerns and place public use of land on a firmer footing. Even so, and in spite of continuing calls from bodies like the Ramblers' Association, 'SNH has concluded that trying to establish in law a formal right to roam on the open hill would be extremely contentious and, most probably, counter-productive' (SNH 1995).

Plate 9.6   SNH recognised a greater need for open-air recreation than was usually acknowledged but also saw signs of increasing friction with landowners. It called for greater planning and funding to ensure that people felt welcome in the countryside and understood their rights and responsibilities. Courtesy of Countryside Commission

GLOBAL CONCERNS AND CONVENTIONS

This was British nature conservation and amenity in the middle of the 1990s. But was it still 'the most advanced system of nature conservation in the world'? Where did it stand in the global context, after the general greening of the universal conscience that had ushered in the decade?

In Britain, everybody had become aware of environmental fears and threats. That is not to say that we came to understand them, or even necessarily to care about them. But they had made the news. Given that the conservation movement was at that time trying to organise itself into a unified entity, how was it that this gift of public awareness was handed to it, on a plate, like a hundredth birthday cake?

As with many silver linings, it enveloped a black cloud; the black cloud of global threats. Threats only guessed at in the past by the wildest scientific minds. A cloud so black that every right-thinking person gave it at least a second thought. Science fiction became reality.

Changes in the atmosphere were threatening the planet's capability to support life; changes wrought entirely by the human species. Chlorofluorocarbon (CFC) gases, developed in the 1950s and used in refrigeration and as aerosol propellants, were found, only twenty years after their formulation, to be destroying the ozone layer. It is this part of the high atmosphere which protects us from contracting skin cancer through excess exposure to the sun's ultra-violet rays. Before 1970, ozone levels above Antarctica had remained constant; then they began to fall. For entirely different reasons, namely the emission of hydrocarbons for which vehicles are largely to blame, low-level ozone was on the increase, causing loss of visibility, damage to vegetation and a threat to health. Carbon dioxide was responsible for 50 per cent of the greenhouse effect; half of this caused by our blitz on the rain forests and half by industry and the burning of fossil fuels. As a result, world temperatures and sea levels rose. Methane was another potent greenhouse gas, on the increase due to changing agricultural practices and seepage from landfill sites. Rain made acid by industrial pollutants shifted about on the weather systems and killed off plant and animal life in many parts of the world.

The picture of rising sea levels, increased radiation and, ultimately, of a planet Earth incapable of supporting the human race provided an undeniable focus of attention. It succeeded where the conservation of species had never succeeded, by striking that most personal, most self-centred and most powerful chord: the fear that we will not survive. Yet, in a hundred years, the nature conservation movement ought to have been able to convince more people of the threat to our survival. It may not be as imminent now as the break-up of the atmosphere but the destruction of species would inevitably have had the same result. For the layman, it would have been easier perhaps to understand than more recent threats.

So, here on the plate, was the message that could capture the world's attention. It is a message, moreover, on which the whole movement can unite, for it threatens everybody's interests, no matter how personal or selfish. Our very own survival is in the balance and might only be guaranteed by our cautious treatment of the environment . . . starting now. No other argument will do. No other argument will interest enough people; no other argument will appeal to the selfish instincts upon which any conservation ethic is founded. Moral considerations are held neither widely nor firmly enough. Other interests are too limited in scope. The man who watches birds in Britain, the woman who surveys her county flora, the teenager who works on a local nature reserve: all will be conservationist in outlook but primarily for those aspects that provide their pleasure. Their reasons for conservation will not attract sufficient converts. The message must have personal meaning for everyone.

How far it will take us towards a sustainable future is, of course, another matter. The road is blocked by so much duplicity and inertia from so many vested interests and by the prevarication and propaganda of the political world. But the individual is discovering his power in a way that nature conservation has never offered him. CFCs were shredding the ozone layer, so the consumer stopped buying CFC products. It was almost as instantaneous as that; almost overnight, manufacturers had to find alternatives or face their products' remaining on the shelves. The felling and burning of the rain forests is releasing so much carbon dioxide into the atmosphere and depriving so many species of their habitat that alternative woods are being found, and greater thought is being given to sustainable practices. Unnecessary pollutants are disappearing from cleaning products; unnecessary colourings and flavourings from foodstuffs. When there is a scare over unhealthy food derived from diseased animals, consumers know they have the power to change the husbandry of those animals. They should also know that they can force changes in the British clothing industry which, in 1985, imported skins of over 25 million fur-bearing animals, including 12,000 wild cats, 140,000 sea otters, coypu and beavers, and 1.8 million musk-rats and marmots. Up to 85 per cent of these may have been farm-produced but, even so, at least 4 million wild creatures were taken that year for Britain to wear.

The 1994 MORI poll for FoE found 42 per cent of adults in Britain claiming to be green consumers. Sixty-six per cent bought ozone-friendly aerosols; 52 per cent bought goods made from recycled materials; 51 per cent ensured that their toiletries had not been tested on animals and 51 per cent tried to reduce fuel consumption at home. That was quite a slice of the market. Green consumerism achieved, in a couple of years, more than the Animal Liberation Front and the vegetarian movement did in decades for farm animals, and more for our environment as a whole than the nature conservationists have done so far.

Quite rightly, there is now some scepticism of green consumerism, which threatens only to postpone the evil day that must inevitably arise from our continued plundering of the planet. We must do more than postpone it; we must cancel it altogether. The quest for growth, no matter how green, must become a quest for sustainability. Development does not have to mean getting bigger or more costly. It can mean getting better: quality rather than quantity. That message could go hand in hand with conservation. What could be better than that? The science and technology are there and, given the willingness of the consumers, economics are there, too. All it takes now is the political will. Even here, the inertia surely cannot last. There are signs that British politics is being dragged, kicking and fighting, towards the environmental ethic by the greener nations, particularly those in Europe. For, assuredly, we have lost our pole position in the world of conservation. Yet, while the MPs puff and blow in indignation when Europe hauls us over the coals (p. 259), they are acutely aware of the conservationists pushing from behind. In between, the British public, as ever too reserved to stand up and be counted, watch with tacit approval. All the signs are that the new global concerns will, one way or another, force international politics to take a greener view of life. It is sad that Britain is no longer leading the way.

Global threats demand global solutions. Something was salvaged from the wreck of the WCS (p. 189) when it was succeeded on 21 October 1991 by *Caring for the Earth – A Strategy of Sustainable Living*. Launched in sixty countries by IUCN, UNEP and WWF, it catalogued the whole gamut of human misdemeanour towards the environment, including deforestation, soil erosion, pollution and population growth. It set out 130 practical suggestions for improving our performance. In particular, it emphasised the need to raise living standards in the developing countries while reducing the scale of their environmental degradation, in order to show that conservation need not be at variance with development. Its overall message, in fact, was that a healthy environment, assured for the future by sustainable ways of living, was the best guarantee of comfort and satisfied aspiration. Its sponsors were careful not to repeat the earlier mistake of talking down to the audience. 'It is not so much a grown-up version of the 1980 document, as a grown-down one – down to a human level' (WWF 1991).

Particularly in its message of global sustainability based upon local action, this new strategy foreshadowed and influenced the UN Conference on Environment and Development (UNCED) in Rio de Janeiro in June 1992. The conference centre, built specially for the event on a drained marshland, hosted leaders from 178 nations who, for a few days at least, turned their minds dutifully to the question of the environment ... and maybe to the growing clamour of the green movement back in their respective countries. The Earth Summit, as it came to be known, drew up conventions on biodiversity and on climatic change, and principles of sustainable development and of forestry. It called upon all participating nations to draft their own Agenda 21 – a series of local and national programmes of action towards sustainability in all things in the twenty-first century.

To those who had taken seriously the findings of the Stockholm conference (p. 137) and the Brundtland Commission (p. 192) and who knew the principles of the WCS and its successor, these ideas of environmental responsibility were nothing new. But, as we have seen, few had taken them seriously. Here was another chance.

One hundred and fifty nations signed the conventions with varying degrees of enthusiasm and reluctance and, in the heat of the moment, one might have thought that the world had finally turned green. But each convention had to be ratified by a minimum of thirty nations before it could come into force and, of course, the energy of Rio rapidly dissipated as heads of government returned to more familiar problems at home. Positive action proved chimerical, but perhaps we should not be surprised; the conservationists themselves were still confused by the scale of these proposals.

Alongside UNCED, the non-government organisations ran their own Global Forum in Rio, though not on a former wetland. Tens of thousands of representatives from a myriad bodies in the movement stole a little of the limelight from the world leaders and gained some credible exposure in the media. For all the euphoria, however, when it came to discussing practical action, it was a salutary experience for many to discover how wide some of the gaps were between themselves and similar groups in other countries. No common policies were found – not even on countering political apathy, let alone on achieving sustainability.

Britain managed more than most in the afterglow of Rio. January 1994 saw the publication of the *UK Biodiversity Action Plan, Strategy for Sustainable Development* and *Sustainable Forestry – The UK Programme*. For eighteen months, the voluntary bodies had fought hard to ensure that these pronouncements would be meaningful, and they were quick now to point out their inadequacies. The main concern was that all the responsibility for these measures would apparently fall upon the official conservation agencies, as if they could function without the active support of all government departments, all organisations and businesses, all individuals. The point was well made for, when a House of Lords Select Committee reviewed the *Strategy for Sustainable Development* in 1995, it reiterated the need for all government departments to take a lead. It particularly slated the Department of Transport and MAFF for their apparent reluctance, and complained that the Treasury was failing to make the polluter pay and to reward innovations in cleaner technology. For all that, it was unable to recommend specific targets.

Similarly, when the signatories to the climate convention met in Berlin in April 1995 to assess their progress towards stabilising emissions of carbon dioxide at 1990 levels by the year 2000, they could only agree that they were failing. Yet they set no binding timetables and merely entreated the 'developed' nations to *prepare* reduction targets by 1997.

The problem for us all was that these strategies embraced a million positive steps that might be taken. They called for a complete review of our attitudes towards the planet, our approach to life. That, of course, was their strength, but it was also their potential for failure. Can we afford to fail one more time?

These pandemic strategies must be kept manageable. Like Agenda 21, they must be broken down ... into continents, into countries, into communities; each and every one of us taking manageable steps towards sustainabilty. Only the voluntary movement with all its components at all levels can hope to keep such a flame alight.

EUROPEAN INITIATIVES

The EC Directive on the Conservation of Natural Habitats and of Wild Fauna and Flora (p. 193) finally came into force on 5 June 1992 and became law in Britain on 30 October 1994. Representing Europe's main contribution to the Biodiversity Convention, like all directives, it was to be implemented by member states by means of their own choice.

The directive listed 169 habitat types and 623 species (over 400 of them plants) worthy of conservation. Birds were not catered for, having their own directive (p. 142). Eighty-two of the habitats and fifty-one species have been recorded at some time in Britain. Twenty-two habitats and one species, the marsh fritillary butterfly, fall within the priority class.

The EC envisaged a network of protected sites, to be known as the 'Natura 2000' series. These would consist of Special Areas of Conservation (SACs) – a new designation – together with the SPAs of the Birds Directive. Member states were to submit details of sites being considered by June 1995; the EC would agree a list by June 1998; full designation should then follow by June 2004.

Aware that the marine environment was especially under-protected, the Habitats Directive required SACs in sub-tidal, intertidal and littoral sites, where fishing, recreation and the discharge of waste would be strictly controlled. It also displayed a new enlightenment by calling for protected corridors and 'stepping stones' between designated sites, and for sustainable management of the countryside in general.

Britain decided that all sites should be given SSSI status in order to attract protection under the Wildlife and Countryside Act. NCO provisions (p. 169) would be strengthened for these Natura 2000 sites. Day-to-day management, however, would continue to be by mutual agreement between all interested parties and, only if this process failed, would government appoint a lead agency to draw up a management scheme. By-laws could be used to give extra protection, but should not interfere with the pertaining rights of any person.

Once again, the theory was riddled with potential problems and the practice quickly faltered. To base the conservation of sites upon SSSI status and the 1981 Act was a joke in itself. SSSI damage and loss continued at a rate that fairly reflected their lack of real protection (Table 9.2). Between 1988 and 1995, 1,005 cases of loss or damage to 3,800 SSSIs had brought about only nine 'prosecutions'. EN was endeavouring to address the problem with its Wildlife Enhancement Scheme, which funded positive management work by SSSI landowners, and its Reserves Enhancement Scheme, which did the same for voluntary bodies on their SSSI reserves. The schemes looked set to spread to most English SSSIs and to be mirrored in other parts of Britain. But, based on voluntary participation and being potentially expensive, they looked unlikely to stop the damage. FoE was concerned enough to promote a Wildlife Bill that would introduce real protection for SSSIs.

Meanwhile, the official agencies were drafting the SAC proposal list without any outside consultation! Apparently the government did not require mutual agreement in all things. They sought only ten sites for each priority habitat and three for non-priority types. Most mathematicians would make that a possible 400 sites; the official proposals listed only 280. Ever since Huxley's list of potential NNRs half a century earlier

Table 9.2   Damage to SSSIs 1990–4

Year	Farming activity		Forestry activity		Planned development		Statutory undertakings	
	(Cases)	(ha)	(Cases)	(ha)	(Cases)	(ha)	(Cases)	(ha)
1990–1	111	70,047	5	12	12	1,323	18	4,240
1991–2	98	30,257	7	39	14	380	19	79
1992–3	50	22,060	6	68	10	64	24	320
1993–4	45	13,826	4	55	13	32	14	670

Year	Recreational		Miscellaneous		Poor Management		Total No.	Total area
	(Cases)	(ha)	(Cases)	(ha)	(Cases)	(ha)	(Sites)	(ha)
1990–1	44	908	69	7,520	12	410	253	84,309
1991–2	28	87	77	4,562	41	1,561	250	36,995
1992–3	16	2,601	53	1,380	33	757	179	27,251
1993–4	23	221	80	2,035	25	1,074	161	16,844

The source for these figures (JNCC *Annual Report 1993–4)* points out that they 'focus on single events such as the building of a road or the drainage of a field. They are much less effective at monitoring the loss of nature conservation interest through long-term processes such as air pollution or lack of suitable land management, notably over-grazing in the uplands. They also concentrate on the negative side of the story. The sharp losses are occurring on a small minority of sites. The majority are in good health and are well managed . . .'

(p. 71), conservationists had called for all components of a series to be protected. Their lists were a minimum requirement, not options from which a selection could be plucked. All examples of those habitats, all sites of threatened species needed designation. Yet, in mid-1995, eight habitats and twenty-two species remained *entirely* ignored. This was largely because to cater for them would require the notification of new SSSIs. This in itself says much about the effectiveness of a designation that was, by then, 46 years old. Corridors and stepping stones and the countryside in general had also failed to attract attention. Much work had yet to be done if Britain were to play its community part in the Habitats Directive.

The EC also reviewed the Common Agricultural Policy (CAP) that was having such far-reaching effects upon our landscape and wildlife. The set-aside scheme had failed to be the environmental saviour that some had expected it to be. There was no enduring advantage to wildlife in setting aside different plots of land on a rotational basis. This simplistic approach had to mature into something meaningful. 'The Set-aside scheme has provided only very limited benefits for nature conservation and we now regard it as a missed environmental opportunity' (NCC 1991).

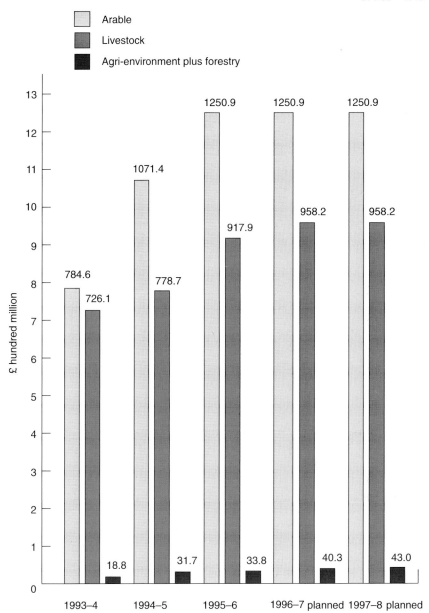

Figure 9.3 Common Agricultural Policy expenditure in the UK

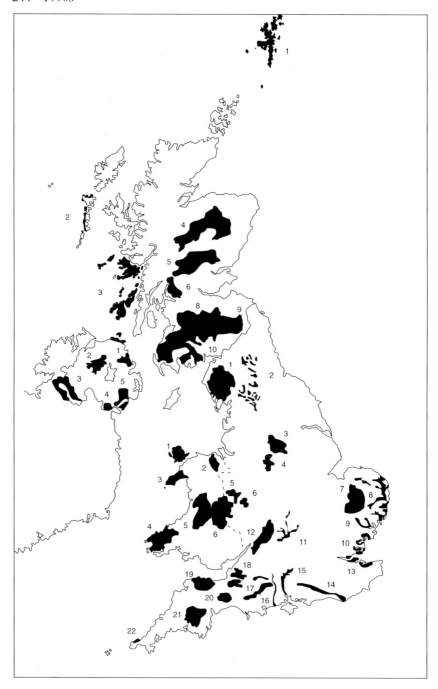

Figure 9.4   Environmentally Sensitive Areas, 1995
Scotland: 1 Shetland Islands; 2 Machair of the Uists and Benbecula, Barra and
Vatersay; 3 Argyll Islands; 4 Cairngorms Straths; 5 Breadalbane; 6 Loch
Lomond; 7 Western Southern Uplands; 8 Central Southern Uplands; 9 Central
Borders; 10 Stewartry

The CAP review of May 1992 took a tentative step away from the production-at-all-costs mentality and towards a more sustainable attitude. Its Agri-Environment Regulation stated an intention to protect and enhance the environment through revised policies, although it fell short of specifying particular objectives and it failed to provide extra funds, so immediate changes were bound to be modest. Nevertheless, here was the EC encouraging farmers – not all of them reluctant to leave the treadmill – to take a more environmental view of their activities. Set-aside land could now become permanent in order to encourage some re-establishment of lost habitats, and schemes to encourage wildlife and public access could be expanded.

In Britain, ESAs were already up and running (p. 175). By March 1994, they extended to 3.35 million ha, including 10 per cent of agricultural land in England and 20 per cent in Scotland (Fig. 9.4).

In England, Countryside Stewardship was also in place. Anticipating, perhaps, the complexion of the forthcoming CAP review, government had tasked the CC to pilot a Countryside Premium initiative in seven counties of eastern England. In seeking to manage set-aside land for environmental benefits, the scheme had succeeded in showing great promise of positive change both in farming practice and in attitudes. It grew into the fully-fledged Countryside Stewardship programme, which was launched in all regions of England in June 1991. The aim was to combine commercial farming and land management with various conservation and amenity objectives in order to:

- sustain the beauty and diversity of landscape,
- improve and extend wildlife habitats,
- conserve archaeological sites and historic features,
- improve opportunities for enjoying the countryside,
- restore neglected land or landscape features,
- create new wildlife habitats and landscapes.

Originally confined to five landscape types, Stewardship was extended as its potential become clear (Fig. 9.5). Work on Community Forests and on farmland at the urban fringe came to attract funds. In 1994, Stewardship

Northern Ireland: 1 Antrim Coast, Glens and Rathlin; 2 Sperrins; 3 West Fermanagh and Erne Lakeland; 4 Slieve Gullion; 5 Mournes and Slieve Croob
England: 1 Lake District; 2 Pennine Dales; 3 North Peak; 4 South West Peak; 5 Clun; 6 Shropshire Hills; 7 Breckland; 8 Broads; 9 Suffolk River Valleys; 10 Essex Coast; 11 Upper Thames Tributaries; 12 Cotswold Hills; 13 North Kent Marshes; 14 South Downs; 15 Test Valley; 16 Avon Valley; 17 South Wessex Downs; 18 Somerset Levels and Moors; 19 Exmoor; 20 Blackdown Hills; 21 Dartmoor; 22 West Penwith
Wales: 1 Anglesey; 2 Clwydian Range; 3 Lleyn Peninsula; 4 Preseli; 5 Cambrian Mountains; 6 Radnor

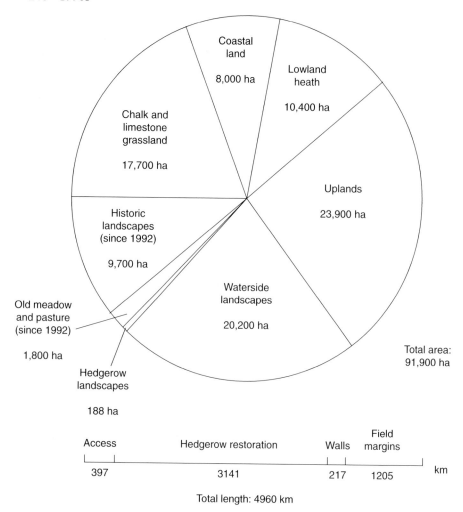

Figure 9.5   Uptake of Countryside Stewardship, 1991–5 (England only)

swallowed up the Hedgerow Incentive Scheme, which had been started two years earlier in the hope of slowing the persistent decline of this vital feature (see Table 7.3). Annual losses for England and Wales were still averaging 18,100 km p.a. in 1990–3, one-fifth by direct removal, the rest through neglect (ITE 1994).

With the Agri-Environment Regulation in place, new environmental and access ventures could be launched. The Habitat Scheme gave payments for the ongoing management for wildlife of former five-year set-aside land. It also supported the creation of saltmarsh on suitable coastal zones, and the enhancement of water-fringe habitats in six pilot areas.

Thirty-two regions were designated nitrate sensitive areas, where financial help was available to stabilise and reduce nitrate levels by lowering inputs to arable land, by the extensification of grassland management, or by the conversion of arable land to grass. In this way, it was hoped that residues in drinking water would be reduced.

The Farm and Conservation Grant Scheme allowed for more traditional management of features such as hedgerows, trees and heather moorland, while a scheme aimed specifically at the moorland habitat was introduced in March 1995. The Farm Woodland Premium Scheme for new plantings was in addition to FC grants. The Organic Aid Scheme sought to increase the 18,590 ha of England registered as being organic in July 1994.

It soon became clear that many of these new schemes would quickly become integrated into Countryside Stewardship. A review of its first three years (MAFF/DoE 1995) showed beyond doubt that it was having positive effects. Ninety-two per cent of access agreements had initially enhanced public access and 88 per cent were continuing to do so. Sixty-three per cent of landscape initiatives were felt to have brought positive change, while 17 per cent had detracted from landscape value. Sixty-seven per cent of projects to enhance wildlife potential had succeeded; 8 per cent had failed. Some of the failures, of course, could be put down to a lack of experience which is perhaps inherent in any new enterprise, and the review called for greater attention to management plans and ready availability of specialist advice and training for landowners.

But for the impending changes to CAP, MAFF would have taken on Countryside Stewardship from the outset. As it was, the CC retained the administrative function, in close liaison with MAFF, EN and English Heritage, until MAFF took over the reins on 1 April 1996, by which time well over five thousand projects were funded at a cost of £11.7 million p.a. Similar schemes of 'stewardship', targeted at their own regional priorities, were under way in Scotland, Wales and Northern Ireland.

This new direction in the countryside, by which farmers with a business to run were able to make allowances for wildlife and public access if they wished, looked set to make environmentally sensitive land use a real possibility in the twenty-first century. But, because it was so novel a direction, it had not yet become integrated into the deeper farming scene. In 1994, £27.1 million was spent on ESA and other environmental measures in British agriculture. This compared with £174 million on simple set-aside and £884 million in arable support. Hardly surprising, perhaps, when set-aside payments averaged out at £45.39 per ha and environmental measures at £2.31. The conservationists called for 10 per cent of all agricultural funding to be aimed at environmental initiatives by 1997. MAFF agreed that greater funding was desirable, but wanted the EC to provide it.

Furthermore, many of the schemes were too narrow in outlook. Pastureland was well catered for, but arable practices deserved scrutiny too. Many were limited to certain areas; while these areas were perhaps more urgently in need of attention, it went against the spirit of the Agri-Environment Regulation to deprive some landowners of opportunities – each proposal ought to be decided upon its own merits. In 1995, the official countryside agencies performed their own review of the CAP reforms and pointed out the need to reduce overall levels of guaranteed spending. Future directions in farming, they felt, required a deeper underlying commitment to environmental and social objectives, based less firmly on financial incentives.

The early 1990s were a time of change in farming attitudes and activities. Wildlife and amenity appeared to have put a foot back in the door after fifty years of exile. The EC had been the catalyst and financier.

The EC, in fact, was still far greener than individual European governments (p. 138) and, when there was talk in 1992 of repatriating environmental policy back to the member states, the voluntary movement, through the EEB, vehemently and successfully opposed such a potentially disastrous move. Late in 1992, the EC Environmental Information Regulation came into force, requiring all but the most sensitive government-held environmental information to be made available to the public. Its Financial Instrument on the Environment (LIFE) provided funds for relevant projects including, but not exclusive to, steps to implement its own directives. In 1993, British projects attracted 2.3 million ECUs in this way. Its fifth environmental action plan, *Towards Sustainability*, was adopted in 1993. Unlike its predecessors (p. 137), it viewed the environment, not as something separate, but as an integral part of the healthy continuation of human activity and development.

The EC was also considering citizen suits, whereby individuals or organisations can take legal action against those who damage the environment.

> The right of citizen suit is based on the concept that a decent environment, pure water, clean air and unspoilt landscapes belong to the community at large, and should not be damaged or taken away for individual profit. It is the principle of 'public trust doctrine' that exists in Anglo-Saxon law, and allows the public to act in defence of the environment, which it has received as a sacred trust.
>
> (Ellington and Burke 1981)

The UK government came out strongly against such an instrument. Feeling that its usual calls for voluntary restraint would not be heeded, it called instead for regulation which historically has been employed to reduce damage until education and persuasion have caught up. But general attitudes towards the environment are trailing even further behind reality than the legislation that is supposed to protect it. The crazy world we have created relies more than ever upon legal pronouncements to settle our differences of opinion and so it seems likely that citizen suits may yet become a reality, as they are in the USA. But only by changing our fundamental attitudes towards the planet can we really expect to survive. Whatever laws we may invoke, that task of education is the greatest challenge of our times.

10 Future: The mechanics and the mission

THE STATE OF PLAY TODAY

There is a conservation crisis and, whether we like it or not, today's generations are not only the first to be deeply involved in it, but the only ones that can take effective action. If we fail to do what is required of us, there will be little that future generations can do, because so much of the raw material will have gone forever.

(Moore 1987)

It is clear where the conservation movement now stands. As a piece of self-perpetuating machinery, it has become highly successful and sophisticated but, as a means of ringing the changes in official actions and attitudes towards our natural heritage, it remains largely ineffectual. The government's official body was dismembered when it made too many demands and 'the radical ecology movement has become "assimilated" with other protests, as a respectable movement with limited non-revolutionary ideas and demands' (Pepper 1984).

The leading lights of the movement are big business, producing glossy magazines, running high-powered sales and advertising departments and generally consuming a fair pool of resources. In building their bridges with industry, if only by accepting its sponsorship, they tacitly endorse its activities, even the damaging ones. Only FoE refuses the assistance of polluters and demands an environmental audit of potential sponsors. In spearheading the movement, the societies have marketed themselves to success. They perform sterling work at a practical level in redeeming some of our past mistakes and offering hope for the future in the form of more positive objectives. But their fortune in influencing the official policies of our administrators has been limited, to say the least. As Lowe and Goyder (1983) have stated: 'their appeal seems escapist' and 'for major reforms ... they will have to bide their time until the resurgence of environmentalism as a social movement'. That resurgence must be their continuing aim whilst, at the same time, they hold the fort by conserving what they can from immediate damage. The conservation movement is our best, almost our only hope for the future, and its success in promoting itself to its present position is all to the good.

> There is a widespread appreciation of the environment and the threat it faces. Of the British adult population, approximately one person in ten belongs to an environmental group. With an estimated two-and-a-half to three million supporters, the environmental movement is now larger than any political party or trade union; its present strength is roughly double that in 1970, which in turn was probably double that of ten years earlier. Simply by virtue of its size and recent growth, the environmental movement qualifies as a major social phenomenon.
>
> (Lowe and Goyder 1983)

A major social phenomenon, but one which is betrayed by its own impotence to alter fundamental outlooks. Larger than any political party or trade union, but receiving little more than lip-service from the politicians.

Let us not make the assumption that even a strong movement would automatically put things right. We must not expect, because the number of followers has grown, that attitudes will change overnight ... or ever. Conservationists are still in the minority; actively committed ones even more so. Their best endeavours have been regularly snubbed. The Wildlife and Countryside Act was a failure; the World Conservation Strategy a non-starter; European Nature Conservation Year a non-event. As William Wilkinson of the NCC put it in 1985: 'nature conservation has recently, despite a spirited rearguard action, been one long retreat'.

So where must the movement take the cause as it enters its second century? How can it ensure that environmentalism does become a social movement?

The vestiges of the official bodies valiantly persevere despite the spells of uncertainty and the changing goalposts. But they, after all, are only the pawns of government, applying the policies that the voluntary sector succeed in pushing through. The voluntary movement must now strengthen its hand and shake off its traditional weaknesses. Certainly, it is beginning to get its house in order, with signs of inner unification and outer universalisation. But will it ever become sufficiently unified and realistic to publish its party line on the really difficult questions?

The time for fudging has passed. Just where does it stand on human population levels, on nuclear power or the use of non-renewable resources? Is a tidal barrage a source of endless clean energy or the ruination of a valuable habitat? Is a wind farm a guarantee for the future or a blight on the landscape? Are the Green Belts places for amenity and conservation or an expensive luxury that pushes development deeper into the countryside? What should we think when a conservationist questions the use of wood-burning stoves? How much public access to the countryside does the naturalist welcome? Choices have to be made. Just what on earth does this 'movement' want?

Clarity and consistency have never been hallmarks of man's attitude to Nature, and they are qualities more in demand than ever as our understanding grows.

(Bonham-Carter 1971)

Why conservationists have lacked clarity and consistency was outlined in Chapter 1. Too many individual reasons based upon personal aesthetics and emotions conspired together to choke consistency. Too many selfish interests, too limited in outlook, left every practitioner with a different aim. Too many feelings, barely guessed, gave clarity no hope within the cause of conservation.

With the 1990s came a subtle change. Whole new problems hit the headlines, more immediately devastating than the loss of species or habitat. Suddenly, the environment was on everybody's mind, not for the sake of the other species but for the sake of our own. Was it the dawn of a 'Green Age'? Environment: discussed in the streets, commented upon in the media, considered by Parliament. Thousands of people employed in a growth industry as countryside advisors and environmental consultants. Other professions taking note of their concerns. Local and national authorities at real pains to appear 'green'.

Conservation was no longer a cranky business; 'green' was go-ahead. In 1993, 88 per cent of people in Britain were concerned about the environment; 51 per cent were very concerned. And only 19 per cent thought that the government was doing enough about it. In 1994, 73 per cent felt affected daily by pollution and environmental damage. Two-thirds of people would have put the environment before the economy. Forty per cent of industrialists agreed that British companies gave too little attention to the environment (MORI poll for FoE).

Quite apart from the nature conservation angle, there was a self-evident amenity need for good, clean countryside. The 1993 UK Day Visits Survey, run by the CC and various official agencies, estimated an annual 590 million recreational trips to the countryside, in the course of which 220 million walks were taken. More than 40 per cent of people made at least one trip per fortnight.

All these people were potential supporters while their attention was kept focused on environmental issues. But the cause had to vie with other concerns: most notably, continuing economic recession. Since development is generally damaging to the environment, recession tends to spell a period of respite from the worst of the threats. At the same time, though, more material social concerns tend to take precedence, with the result that conservation activity also stagnates. Official funding may be frozen and voluntary bodies see a slump in membership growth.

Many organisations were struggling to maintain membership figures and resources at 1990 levels (See Table 8.2, Fig. 10.1). Those that succeeded

Plate 10.1 Over 40 per cent of people make at least one trip into the country-side each fortnight. They should all be encouraged as potential supporters of the cause: Countryside Commission Interpretation Centre. Photograph by Mark Boulton. Courtesy of Countryside Commission

did so only by working very much harder at it. Particularly telling, though, is the fact that the direct action groups maintained a healthy growth.

It is a Catch 22 situation for the movement. Economic revival may assist conservationist activity but it will also increase the amount of work to be done. On the other hand, continued recession will see the movement progressing at a much reduced pace. Lowe and Goyder (1983) pick out a ray of hope, however, in the event of long periods of recession.

> long-term changes in the structure of employment – such as the continued shift towards service occupations – could contribute to a resurgence of environmentalism. Moreover, the experience of unem-ployment over an extended period of time may, as in the 1930s, turn people away from the workplace as a source of fulfilment. Resigned to the absence of work, they may look to non-economic spheres for self-fulfilment and political expression, and begin to exhibit some of the post-material values, including concern for environmental conservation.

Ultimately, however, instead of depending upon the rather problematical course of economic politics, the movement should be aiming to influence politics in its own right. For, while recession could be blamed for stagnating

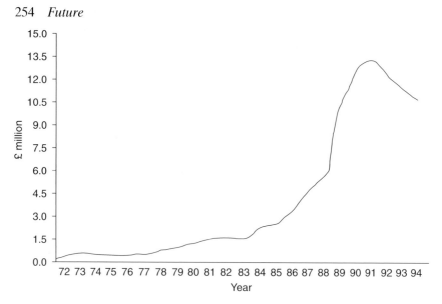

Figure 10.1 World Wide Fund for Nature (UK) expenditure on projects worldwide

income, it did not account for the subtle loss of urgency that had crept into environmental affairs. WWF (1995) was not alone in seeing the 1980s as the time when 'the environment was at the top of people's minds'. The 1994 MORI poll concluded that environmental concerns had lost their edge since the start of the decade. In 1988, 14 per cent of adults would have written to their MP about environmental matters; in 1991, 31 per cent; 1992, 23 per cent and 1994, 26 per cent. The movement needed to work very much harder at re-igniting the urgency as well as the membership tally.

SOCIAL CONSERVATION

The pool of potential support was growing all the time. The traditional camp has been that middle section of the population that has achieved a comfortable style of living and has a little left over to indulge its leisure. These are the people who have felt an improved quality of life through a passing acquaintance with nature and the countryside. Their ranks are swelling as general affluence increases. But the recent concerns for the future of the planet have transcended these traditional lines and attracted interest from all sections of society. Thus the less privileged, previously concerned with the day-to-day problems of living, now find time to consider the environment. Those more privileged, whose wealth and positions have historically cushioned them from the worst problems of the world, can no longer afford to ignore the prophecies. Increased time for leisure, greater mobility, improved wages and education all encourage calls for a better world to live in. The younger generations are proving more altruistic in outlook as social values change. The highly motivated individual, to whom leisure is anathema and 'success' an end to justify any means, is being regarded in society with increasing suspicion.

There is also a greater willingness to act; to dirty the hands and shout the odds ... and not always for payment. Much of this emanates from the type of delegating government that Britain has enjoyed or endured since the start of the 1980s. Responsibility – and payment – for action has increasingly shifted from central government onto the shoulders of society and the individual. It is a phenomenon upon which the movement can capitalise. Indeed, the movement has always made good use of the volunteer. We have talked about the feeling of individual helplessness to assist the cause. We also talk about the loss of personal esteem through unemployment. The conservation movement has an almost infinite capacity to occupy the volunteer.

In the past, there has tended to be a suspicion of voluntary work; to many people not involved in it, it was itself an integral element of fringe-group activity. It reeked of an unhealthy independence of thought and action and was an uncomfortable element in an organised society. But attitudes have softened. An enlightened nation of educated people can be allowed the freedom to express themselves through voluntary work. In a free and enquiring society, it is only right that individuals be given the opportunity to do something for a cause which they support. At a practical level, the easing of personal concerns over health and financial hardship and the vast increase in general leisure time have made it possible, even desirable, to take up voluntary work.

> the role of the volunteer is changing and will continue to change in so dramatic a way as to make the traditional meaning of the word largely irrelevant. 'Volunteering' is fast ceasing to be an activity practised by a small minority for the benefit of the majority but is becoming the natural means whereby the majority of citizens may become involved in their own community, whether in the form of pressure groups, whether by physical work in their immediate environment or in other ways. The distinction between 'voluntary work' as a minority service function and 'voluntary work' as the natural expression of individuals' desire to be involved in their own society ... is a critical assumption.
>
> (Stevenson 1972)

In 1994, the BTCV achieved 112,400 volunteer work days. Its five-year-old International Conservation Action Network consisted of over 180 organisations in forty-two countries. Twenty-nine thousand volunteers gave the National Trust 1.7 million hours in 1994–5, the equivalent of one thousand full-time staff.

By involving an ever-widening circle of interests and by teasing out individual personal commitment to the cause, the conservationists will eventually attract the attention of the politicians. For it is the nature of politics to react to the public mood; to give precedence to points of greatest public concern. Throughout this book, there have been examples of political failure to recognise the conservationist argument. But, of course, the failure has not been that of the politicians; it has been the failure of the conservation movement itself, which has proved consistently unable to mobilise widespread concern among the British people. Conservation remains a painfully minority interest and, until it begins to hold more sway than the farmers, than big business, than the road transport lobby, political decisions will continue to go against it. For, until now, all official conservationist activity has been moulded by that omnipotent force – economics.

There were signs in the early 1990s, however, that politics and business were having to take account of this new willingness to act, particularly in local people fighting for their surroundings. The road building programme, for instance, was drastically modified and, whatever the official reasons given for it, there is no doubt that the vehemence of some campaigns by local residents stuck in the minds of the politicians. Local groups were the backbone of an alliance that prevented a road being put through 240 homes and the eight-thousand-year old Oxleas oakwood SSSI in South London. Eight years of legal wrangling, direct action, two public enquiries and the involvement of national bodies, the High Court and the EC finally ended when the plans were revoked in July 1993. An eight-year campaign against the M3 extension across Twyford Down in Hampshire had failed in the previous year, but not without forcing politician and developer alike to review their attitudes towards public opinion. In 1994, the government scrapped fifty-three of its many road schemes, put seventy-eight on hold, and altered plans for a fourteen-lane M25 'superhighway'. Further pruning of the programme was announced in the 1995 autumn budget.

This was selfish conservation at its most fundamental and most potent. Here were local issues being raised in the national conscience and bringing about change in national policy. Here was government at last listening to individuals and not to the industries that would make a quick fortune and pull out. Democracy works only when enough people get up to shout the odds or get down to dirty their hands. In this day and age, the local environment appears to be one of the most potent catalysts.

As well as the 'grass roots' campaigns, there is also a place in the movement for high-profile, direct action, as practised by FoE and Greenpeace. The giant multi-national company, Shell, planned to dump the Brent Spar oil platform in the North Atlantic. It did its homework. All things considered – including cost – this seemed the best option. Greenpeace members camped out on the platform and so made the plans newsworthy. Greenpeace did its homework and came up with a different, and inaccurate, set of figures. Shell knew that Greenpeace had it wrong, but nevertheless backed down in the face of incensed public opinion and a threatened consumer boycott of its products. The public were ignoring all the hard-wrought figures; the fact was that it just did not seem the right thing to do. Nothing Shell was saying could alter that gut feeling. The Precautionary Principle – 'Take care' – won the day.

While it may have been novel to see industry failing to court public opinion, the conservation movement was painfully experienced in not getting its messages across. It 'won' Brent Spar for the wrong reasons. Too often in the past it had not done its homework, had failed to come up with the facts. But the scaremongering should have gone out with the 1960s. After the flamboyant direct action has finished, and the adrenalin subsides, the movement must show a firm factual base for its opinions.

> Human needs, immediate and future, of the flesh and of the spirit, are being severely compromised by our lack of understanding of how nature works and where we fit into its patterns. At present, the fabric of the natural world that supports us is rotting away thread by thread because of the way we work and intend to develop.
>
> (Durrell 1986)

Urgent and high-profile campaigns have brought some swift reaction from the policy-makers in the past but have not always been based on scientific truths. The search for those truths has often been too slow and deliberate for the action to wait. But future campaigners must have access to the facts. Scientists disagree over the exact nature of the greenhouse effect and the threat of global warming. Concern has been expressed over the aerosol propellants that replaced the CFCs. The plight of the seals (p. 186) had long been forgotten by the media before a cause of the virus could be given. Much research is performed or funded by biased bodies with a vested interest in a certain result. The financing and management of environmental research must be raised to a realistic level so that voluntary bodies, amateur and professional naturalists, government, developers; all may help to come up with the correct information. So much work needs to be done. Too many arguments are being lost through the lack of facts and figures. It is barely more acceptable to win them for the same reason.

Scientific research is a costly business. So are land acquisition and management, the introduction and enforcement of laws ... and all the other components of conservation. Many organisations are finding how expensive it can be merely to retain current members. Where is the money to come from?

Figure 10.2 shows government funding for conservation and amenity to be faltering even at current levels. Compare the money available in 1994–5 to MAFF (£2,810 million), DoE (£1,980 million) or Department of Transport (£5,850 million). There we have the picture of conservation's place in the British political game. But even reduced funding for the official agencies need not necessarily mean that overall resources are cut.

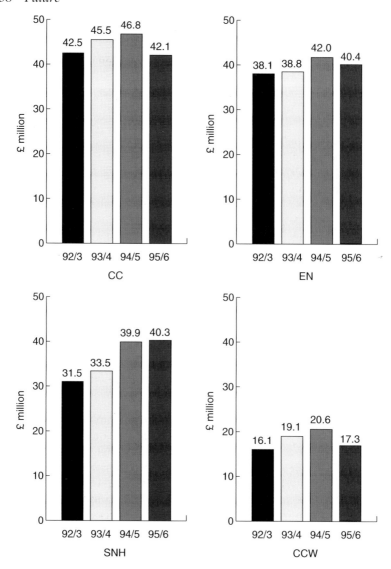

Figure 10.2   Government Grant-in-aid to conservation and amenity

What will be most vital is the redirection of other funds towards environmental initiatives, such as Countryside Stewardship. In that way, the conservationist ethic may begin to suffuse all our activities. It is part of that political process of encouraging society in general to do those things that government would once have been expected to take on board. In this, the voluntary conservation movement can play a more vital role than ever. It must demand realistic levels of funding, from whatever source; it must ensure that it reaches the right parts; it must be on hand with expertise and experience to see that it is wisely used.

Increasingly, finance is likely to come from the generally greener EC. It currently funds many of the environmental initiatives in agriculture and will undoubtedly lead the way in 'greening' forestry and fishing, industry and commerce, tourism and all forms of development. This revenue will be welcomed, even by our island race, for most Britons, either openly or tacitly, accept Europe's keener environmental outlook. In the early 1990s, time and time again, the EC hauled Britain over the coals for its poor showing. Legal proceedings were started over the road-building schemes at Oxleas Wood and Twyford Down. Action was taken over the lack of legislation on nitrogen oxide emissions. Proceedings were threatened over the tardiness of SPA designations. On each occasion, as the UK government acted scandalised that Europe should have the temerity to find fault with it, British public opinion gladly backed the continent. The reason is plain enough.

As the concerns of the movement become global in nature, what are national policies but isolated, local efforts? It is all a question of scale. If Britain's policy-makers take their precious environment rather too much for granted, they can be sure the local residents will be there to chivvy them along. They must develop the sort of concern that we all feel for our local surroundings. That will mean a part of the movement playing them at their own political game and relentlessly chipping away at their outmoded attitudes.

> the conservation movement will never achieve its objectives if it tries to rely entirely on goodwill and persuasion, evades the conflicts of interest and denies the political nature of the issues it is raising. Conflict raised to the level of violence is destructive. But conflict within the limits of democratic debate and decision-making is a creative force without which society ossifies. The future of the countryside may ultimately depend on whether it remains a political backwater in which the ideology and interests of the rural establishment go unchallenged, or whether it becomes a focus of political debate and action.
>
> (MacEwen 1976)

While conservation struggles to become an issue in British national politics, we shall continue to rely upon Europe to lead the way in environmental action and funding, for, unlike thirty years ago, Britain quite demonstrably no longer has 'the most advanced system of nature conservation in the world'.

The movement must be commended, though, for keeping the flame alive. That in itself has been no mean feat. In one hundred years, it has proved resilient enough to survive an age when 'environment' was an unknown word and nature study was a pastime to be mocked. But, like the environment, the movement must remain dynamic. Public attitudes

Plate 10.2   Which direction to take in the future? We ask so much of the countryside and of the movement that seeks to conserve it. Top priority must be for conservation to shake off its aura of negativity and become a positive force. Photograph by Simon Warner. Courtesy of Countryside Commission

– and, yes, political ones too – are open to persuasion now that everyone has heard that the planet is sick. There has never been a better opportunity to put the ethic of sustainability at the top of our daily agenda – to bring about an attitude of automatic, instinctive conservation that treats the entire world as a nature reserve, a place for recreation, a resource for the future. Such an outcome might be achieved by legislation or education, but it is by no means certain that it will be achieved at all. In which case, the human race will be leaving behind a very much poorer planet.

Will the movement itself be able to adapt to the new circumstances? Can it accept that our species will continue to make great demands upon the environment? Perhaps by promoting sustainability in all spheres, it may at least throw off its mantle of negativity, for this recent doctrine preaches that development need not necessarily prove inimical to nature. SNH is promoting 'Targeted Inputs for a Better Rural Environment' to show how technology – one traditional enemy of nature – can encourage intensive farming – another one – to become more environmentally conscious. It is one example of many new initiatives being put forward by the movement. Is conservation about to become a positive force in the British countryside?

How will the movement handle the realities of conservation on the global scale? The day will come when Britain must lose a habitat of national importance for the sake of saving one of international value. Will the conservationist attitude cope with such a global choice? There will be animals and plants lost from our shores that will thrive in number elsewhere, while we continue to care for those with strongholds in Britain. Will the island race counter such international conservation? The world is so small that one nation's problems come to concern us all. Acid rain, the destruction of tropical forests, the fraying of the ozone layer; they transcend national boundaries and concern every human being on earth. How will a movement founded on self-centred motivations, which has known but limited success on a national scale, prosper on the international scene?

UNNATURAL NATURE

The practical aspects of nature conservation, as well as the theory, are undergoing continual change. As the wider countryside and the whole environment fall within its domain, the bastions of nature reserves will assume a lesser role in the overall masterplan. Deleterious effects upon the environment in general infiltrate the most closely protected reserve. Tags and labels are impotent against pervading pollutants that threaten our plants. Gates and fences cannot keep the animals and birds from going beyond their bounds.

Reserves will still play their part, of course, but we shall see fundamental changes in their administration and management, with a general shift of responsibility to smaller, more localised groups, to volunteers and to private individuals. Many NNRs are subject to leasehold and management agreements which are running out. All the dedication and costly management work will be proved as worthless as the designation itself if the future of these sites cannot be put upon a more secure footing. To compound the confusion of the conservationists, proposals have been made to sell off the freehold NNRs. In this event, the most valuable sites will undoubtedly fall into sympathetic hands, albeit at great cost to the voluntary movement, and enjoy a greater security in the long run. But the fate of the remainder would be far from certain. Many practitioners are sceptical of attempts to sell off a resource which has no definable market value, but they overlook the fact that the voluntary bodies have bought many reserves on the open market. It is no different to putting a price on any other piece of land.

The fear is that marketing will open the floodgates of commercialism on the conservation scene when there are very few people competent to manage our fragile habitats. Yet the attraction of owning one would undoubtedly encourage a heavy inflow of funds and, for the genuine owner, it is only a question of training to overcome all but the most insurmountable of problems. Many owners would employ an experienced 'warden', and only the most intransigent would allow what they had paid for to spoil. In those cases, a restrictive covenant, imposed at the time of sale, could provide the safety net of a supervising body to carry out the necessary work at the owner's expense, as is the case with listed buildings. Since the vast majority of nature reserves today are privately owned, the idea is hardly innovative. The official agencies, due to financial stringencies, already actively encourage the management of some NNRs by approved bodies, such as the voluntary societies, wildlife trusts, local authorities, even industry. Local authorities are also taking on greater responsibility for areas such as SSSIs and country parks.

The most valid concern in all this 'economising' of nature is that we shall too easily forget all the non-monetary value inherent in it.

> We consider that exercises which attempt to place a cash value on wildlife for use in cost-benefit analysis are potentially misleading. Although the latter are intended only as an aid to decision-making, placing an estimated price on wildlife can detract from the less tangible values that society places on wildlife and attractive surroundings, and could easily provide a spurious justification for inappropriate development of SSSIs. The use of such market mechanisms should, we believe, be limited to providing incentives to encourage the right management of SSSIs and the wider countryside.
>
> (NCC 1991)

The official agencies, already heavily committed to paying landowners to conserve, have also been forced to consider 'charging for nature', although there is a 'proper reluctance' to do so. Many of the voluntary bodies charge an entry fee to their reserves, particularly for non-members. The movement finds itself in the grip of another irony: while demanding that we all, now and in the future, retain the right to enjoy the natural world we tenant, it finds its efforts on our behalf increasingly costly. Is it morally right to charge for a right? Will practical necessity prevail in the end?

Again, the answer revolves around attitudes, around conservation's place in society. At present, it retains the charitable status that has haunted its history. When it comes to be seen as the sole guarantee for our continuing presence on Earth, charging for it will become morally acceptable and right. Sponsorship, too, will be more socially commendable than it is at the moment and will increase accordingly. Central government will be forced to fund conservation realistically.

An exciting prospect is the creation of new habitat, the restoration of spoilt habitat and the reinstatement of lost species. Again we can expect much ethic-searching and questioning of principles but such moves merely represent an extension of our continued moulding of the artificial countryside.

New wildlife habitat is less artificial than the parks and gardens which, only now, are being tended more with wildlife in mind. Half a million hectares of private gardens in Britain afford their owners one way of showing concern for nature and bringing them closer to those lost rural roots. They represent an important series of oases for wildlife but, of course, will never be as valuable as what they tamed in the first place. There are one hundred thousand hectares of golf course with unrealised wildlife potential, and thirty thousand hectares of open space around our remaining railways, while disused tracks are made corridors of re-establishment and return.

Roadside verges extend to more than two hundred thousand hectares and increase with every new road built. They have their own flora and fauna, specially adapted to the prevailing conditions. Inappropriate pesticides are now rarely used on them; cutting or mowing is timed to avoid the peak flowering months. Where salting of the road is regular, salt-tolerant species of plant move in, their seeds carried by passing vehicles as well as by more traditional modes of dispersal. Thus, seaside plants can be found on inland motorway verges. In 1995, the Kent Trust for Nature Conservation designated Bench Hill, near Woodchurch, as the first Roadside Nature Reserve in its Road Verge Project to find the ten best verges for wildlife in each of the county's districts.

So robust and adaptable is our wildlife that there have been many gains as well as losses resulting from our manipulations. Consider all the birds that rely on ploughed ground: the stone curlew and skylark, the curlew, lapwing and oystercatcher. Much farmland wildlife is under threat from changing agricultural methods but much of it would never have been there in the first place but for our activities. The great crested grebe looked set to disappear from Britain in the nineteenth century but, thanks to the proliferation of man-made gravel pits, there are now about eight thousand birds. Since 1938, the little ringed plover has colonised gravel pits, sewage farms, reservoirs and tips, reaching a total of some four hundred pairs. The sand martin relies heavily upon gravel pits for potential nest-sites. Some gulls have moved much of their population into towns and cities, where easy food from rubbish tips and breeding sites on tall buildings are readily available. The red fox finds easy pickings in the city. The black redstart, first recorded in Britain in the eighteenth century, began breeding in 1923 on the sea cliffs of Sussex. With the onset of war, it increased its numbers by colonising bombed sites and subsequently moved into industrial areas. Even our most disastrous assignments are turned to good effect.

> There is probably nowhere in Great Britain where the flora is truly wild, in the sense that there is no part of the flora that has been unaffected by man's activities. A lot of our flora is semi-natural, in the sense that it has come of its own accord, but that is not quite the same thing.
> (Moriarty 1975)

Ratcliffe (1977) found that native plant species occurring in natural or semi-natural vegetation may be as few as one thousand in number. But include all species growing in the wild 'even if only ephemerally on a rubbish dump' and the number may well exceed two thousand, or three thousand if certain micro-species are included. Of the 1,700 plant species he considers, 406 (24 per cent) are dependent, in Britain, on man-made habitats or human activity. Of these, 229 are native, 177 are introductions. Two-thirds of them could not survive without human influence, and the remainder would be much reduced in number. Over a tenth of them are national rarities, representing a sixth of Britain's total of 280 rarities.

Of the same 406 species, 66 per cent occur on agricultural land and 94 per cent are found in non-agricultural man-made habitats. Of these habitats, roadside verges and hedges, urban waste ground, building sites and rubbish dumps are the most important. Ratcliffe also points out 'a strong case for the preservation of the flora of the walls of ancient buildings'!

So a substantial proportion of our flora and fauna is heavily dependent upon human activity and thrives most healthily in artificial habitats. The creation of habitat specifically for wildlife – the creation of nature reserves – is but the next stage, a land use as valid as any other. If semi-natural species come of their own accord, they ought to be welcomed, being no less natural than most of what we seek to conserve today. And there may also be a place here for those species that do not fare well under our present regime: the agricultural weeds, for example. 'Traditional meadows' are being developed in urban green space; whole new woods planted; damp field corners reinstated. Endless possibilities present themselves.

Saltmarsh is being recreated on the Blackwater Estuary, and the RSPB is converting 240 ha of arable farmland at Lakenheath Fen in Suffolk back to its original wetland state to form the largest new reedbed reserve in Britain. Rutland Water in Leicestershire, a reservoir built at the end of the 1970s, was acclaimed by Sir David Attenborough as the finest example of creative conservation in Great Britain. According to Ratcliffe (1977), 'artificial waters now form the nearest alternative' to our all-but-lost fenland meres. They have become home to hundreds of wintering wild-fowl, adopted by dragonflies and duckweed, a vital wildlife habitat. Derelict areas of urban blight are being made to bloom and breathe again. Teesside Development Corporation is creating a large nature reserve of reedbed, swamp, water, woodland and meadow on land once owned by ICI. 'The scope for this creative conservation is almost limitless' (NCC 1984).

But, as with any new idea, enthusiasm must be tempered with well-founded caution. The greatest fear is that these new reserves could come to replace the genuine article; that development of important sites would be permitted in the belief that nature could move elsewhere. That, absolutely, is not the case, even though the courts have taken the view that an SSSI can be traded for an equivalent area of totally different land. If that opinion is allowed its head, our wildlife will be pushed from pillar to post until nothing of value remains.

Ethically, too, there is a strong aversion to blatantly artificial 'nature'. Conservation is strong in its aesthetic appeal and it is likely that the following picture will always seem cold, futuristic and unworthy of preservation:

> 'Synthetic' nature reserves, on the lines of some of those in the Netherlands, where artificial habitats have been constructed, would help; some of our rarer and more threatened plants and animals could be introduced into these under careful control by experts. These would not only provide 'biological gene banks' to assist the survival of species but also (as in some zoological nature parks elsewhere) places where the public could see many species, without pressurizing their few relict natural populations. The latter could then be legitimately the subject of much stricter controls.
>
> (Rose 1984)

One of the main reasons for conserving species is to preserve the plethora of genes in the natural world for our future exploitation. Artificial gene banks act as a safety net should any species be wiped out by accident or design. In 1975, the Biological Records Centre at Monks Wood began registering living material of nationally rare plant species. Specimens are made available to scientists so that collection from the wild may be avoided. The Royal Botanic Gardens are establishing a new seed bank where scientists can research the beneficial properties of plants, less than

5 per cent of which have yet been scrutinised. The National Council for the Conservation of Plants and Gardens, with forty county groups and over seven thousand members, encourages the propagation and distribution of rare plants amongst its members and collections. In December 1989, IUCN and WWF launched an international Plants Conservation Programme to 'allow botanic gardens to occupy the leading position in plant conservation and resource development that rightly belongs to them'. Tragically, the conservation of the world's flora could no longer be risked solely in the wider countryside. Necessary though such moves may be, they are aesthetically unacceptable and could undermine one of the strongest foundation blocks of the conservationist ethic.

Plate 10.3 If artificial creatures manufactured in laboratories are not to replace this golden ringed dragonfly, we had better concentrate on recreating some of its wetland habitat. Photograph by Colin Horsman. Courtesy of Countryside Commission

If plants confined to collections seems futuristic, consider the work of genetic engineers who seek to develop artificial creatures or to modify natural viruses for use as insecticides. What of the computerised breeding programme that looks for genetically ideal mating between the world's few remaining Californian condors? Captive breeding gives the birds a twelvefold increased chance of success. Many conservationists vehemently opposed the birds' capture and removal to zoos when it was carried out in 1986. Somehow, even with the condor's best interests at heart, it did not seem ethical. How would British conservationists have reacted to a similar plan to save our relict population of red kites, had the NCC and RSPB not preferred a programme of re-establishment using foreign stock?

The question of extraction, of taking organisms 'into care' from their environment, has only recently posed problems of conscience for the movement. But introductions to the wild, and the associated ideas of re-introduction, reinforcement and translocation, have been topics of controversy for centuries. Many of our best-loved trees were introduced. Vast numbers of wild plants are derived from garden varieties. Many birds and animals are with us only by our own hand: the little owl, red-legged partridge, grey squirrel, sika and muntjac deer, to name but a few.

Arguments abound both for and against artificial establishment. It has sometimes saved endangered species. It can supplement and vary a community. It can be a useful method of research into habitat and population dynamics and, for this, it need not be a permanent establishment. It has been used as a biological means of pest control. Many examples have been purely for human enjoyment.

But dangers exist in putting alien species into an environment. Some exotic species have an appetite for crops or indigenous plants; others simply wreck the environment, such as the coypu which does untold damage to banks and drainage ditches. They may supplant a native species by taking over its habitat, as the grey squirrel has the red. Mink have likewise ousted the otter in some regions. The American signal crayfish has decimated populations of our native Atlantic stream crayfish by carrying a virulent plague with it when it began to escape from the fish farms as soon as it was introduced in the 1970s. Genetic contamination of native populations can result from interbreeding with alien species, as with the white-headed duck in Europe which now cohorts with American ruddy ducks that have escaped from collections and established feral communities.

Such contamination is most worrying within reinforcement programmes, or where the original species is mistakenly believed to be extinct. Where very rare species are concerned, the source population may be jeopardised by the removal of individuals, and serious thought must be given to the high incidence of failure in establishment schemes. Captive or nursery breeding of additional stock must then be considered but this has proved extremely difficult with some species. One of the condors released in

1995 had to be recaptured and trained in the perils of approaching people after it was found eating popcorn and hot-dogs on an August camp-site! Schemes involving butterflies and moths have perhaps been the most successful in this field. It is imperative that all artificial establishment is recorded precisely so as not to cloud the picture of natural distributions.

For all these dangers and concerns, there have been many successful schemes over the centuries and they currently play a prominent role in the various species recovery programmes of the official agencies. Again, it is a matter that calls for agreement and co-ordination between all interested parties. Since the Wildlife and Countryside Act placed a general prohibition on the introduction of alien species, the whims and fancies of individuals have posed less of a threat and any schemes these days are well-prepared and documented. The 1992 Biodiversity Convention called upon all signatories to 'prevent the introduction of, control or eradicate those alien species which threaten ecosystems, habitats or species'. In January 1994, the IUCN began drawing up international guidelines on species manipulation.

Science will often find a way, of course, but the acceptance or otherwise of many of these ideas comes down to the question of ethics and again rests upon the fundamental whys and wherefores of conservation. Few would fault such measures where the aim is to preserve a species or habitat from complete extinction. Most people are comfortable with the re-establishment of an organism in an area from which we have caused its disappearance, particularly if the loss is recent. Many prefer biological pest control to chemical measures, although the threat to our wildlife communities could, in fact, be as great. But how far is it acceptable to interfere with 'natural' populations for the sake of scientific research? As for personal preference or pride, whim or ornament, as many of the early experiments were, most natural historians and conservationists today would never contemplate it.

Thus the movement must decide each case on its merits. While our tinkering may make the already unnatural wildlife a little more or a little less natural, it can also make it incomparably more varied. But:

> it must be remembered that whilst we are concerned about saving the species, if this was our sole concern this could be carried out adequately in botanic gardens [and zoos] and we would not need nature reserves at all. What we are trying to save is the species in its natural community. If we take the attitude that species can be conserved by moving them to other sites where they have never apparently been native we weaken the case for saving the sites where it is native.
>
> (Perring 1975)

THE MISSION

That we can organise and operate a very sophisticated, world-wide financial system and yet not know how to safeguard adequately the natural resources that we and future generations will depend on is a direct challenge to our current attitudes, and a disturbing irony not to be shrugged off.

(Kinnersley 1994)

So many things to think about on a hundredth birthday. So many new ideas to consider; new angles to try out on old problems. Old images to shake off; fresh partnerships to forge. Lots of people asking how they might help; lots more still to convert. More resources needed; more information to distil. Yet no guarantee of a further hundred years in which to achieve it all.

The movement is faced with a mission near-impossible – to change the fundamental outlook of, first, a nation and then the world, from greed and materialism and continued striving upwards to gentle coasting into an infinite and comfortable future.

But is it really such an impossible dream? Need we really live so at variance with the natural cycles and rhythms of our planet? Much of what we seek to preserve in the way of habitat and landscape has resulted from our own activities. The woodlands and the copses, the pastures and the hedgerows, the parks and orchards and many of the lakes – all have been made by the human hand. They are tenanted by animals and plants that can adjust to change if given half a chance. Much of what we have lost can be recovered if we act correctly. There is nothing inevitable about our wasting the environment; with care and cunning, we need not destroy it and, with informed foresight, we shall not wish to.

The spectacle of so many groups of intelligent people doing their utmost to destroy the vital life support systems of their posterity on earth, and angrily opposing efforts by more enlightened groups to mitigate or prevent the mischief, is a truly remarkable phenomenon of our age, and one which we can be sure our successors, if they are permitted to survive on earth, will find it difficult to credit.

(Nicholson 1987)

The conservation movement has many successes behind it, the greatest of which has been its survival and increase against all odds in its formative years. Now that it is established, it must face up to its most vital challenge: to bring conservationist attitudes to all people in all things. Only through greed and shortsightedness have we forgotten our natural inclination to cherish the world. The conservationists have it within their power to re-direct our greed and reopen our eyes. If they fail in this, their history will have all been for nothing. It is their duty to succeed.

References and further reading

Unless otherwise stated, the place of publication is London.

Adams, W.M. (1986) *Nature's Place: Conservation Sites and Countryside Change*, Allen and Unwin.

Addison, C. (1931) *Report of the National Park Committee*, Cmnd 3851, HMSO.

Allen, D.E. (1976) *The Naturalist in Britain: A Social History*, Allen Lane.

Allen, R. (1980) *How To Save The World: Strategy for World Conservation*, Kogan Page.

Ashby, Lord E. (1978) *Reconciling Man with Environment*, Oxford: Oxford University Press.

Baker, J. (1986) *Striking a Balance*, CEGB.

Barr, J. (1969) *Derelict Britain*, Penguin.

Beinart, W. and Coates, P. (1995) *The Taming of Nature*, Routledge.

Blunden, J. and Curry, N. (eds) (1990) *A People's Charter? Forty Years of the National Parks and Access to the Countryside Act 1949*, Cheltenham: CC.

Body, R. (1982) *Agriculture: The Triumph and the Shame*, Aldershot: Temple Smith.

—— (1984) *Farming in the Clouds*, Aldershot: Temple Smith.

Bonham-Carter, V. (1971) *The Survival of the English Countryside*, Hodder and Stoughton.

Boston, R. *et al.*(eds) (1979) *The Little Green Book*, Aldershot: Wildwood House.

Bourassa, S.C. (1991) *The Aesthetics of Landscape*, Belhaven.

British Trust for Conservation Volunteers (1972) *Organising a Local Conservation Corps*, Wallingford: BTCV.

Brown, A. (ed.) (1992) *The UK Environment*, HMSO/DoE.

Bunce, M. (1994) *The Countryside Ideal: Anglo-American Images of Landscape*, Routledge.

Carson, R. (1963) *Silent Spring*, Hamish Hamilton.

Cantell, T. (1977) *Urban Wasteland*, Civic Trust.

Cherry, G. (1975) *National Parks and Recreation in the Countryside*, Cheltenham: CC.

Chisholm, A. (ed.) (1972) *Philosophers of the Earth*, Sidgwick and Jackson.

Coburn, O. (1950) *Youth Hostel Story*, The National Council of Social Service.

Cole, L. (1980) *Wildlife in the City: A Study of Practical Conservation Projects*, Peterborough: NCC.

Commission of the European Communities (1990) *The European Community's Environment Policy*, Brussels: CEC.

—— (1992) *Towards Sustainability – A European Programme of Policy and Action in Relation to the Environment and Sustainable Development*, Brussels: CEC.

Conder, D. (1986) 'Two green forces', *Environment Now*, pre-launch issue, December 1986.

Conservation Foundation (1982) *The First Conservation Annual*, Cambridge: Granta.

—— (1985) *The Conservation Review*, Exeter: Webb and Bower.

Conway, G.R. and Pretty, J.N. (1991) *Unwelcome Harvest: Agriculture and Pollution*, Earthscan.

Council for National Parks (1986) *50 Years for National Parks*, CNP.

Council for the Protection of Rural England (1988) *Liquid Assets*, CPRE.

Countryside Commission (1972) *The Use of Voluntary Labour in the Countryside of England and Wales*, Cheltenham: CC.

—— (1980) *Trends in Tourism and Recreation 1968–78*, Cheltenham: CC.

—— (1987) *Forestry in the Countryside*, Cheltenham: CC.

—— (1987) *New Opportunities for the Countryside*, Cheltenham: CC.

—— (1989) *Paths, routes and trails: Policies and priorities*, Cheltenham: CC.

—— (1991) *Areas of Outstanding Natural Beauty: A policy statement*, Cheltenham: CC.

—— (1991) *Caring for the Countryside: A policy agenda for England in the Nineties*, Cheltenham: CC.

Countryside Commission for Scotland (1978) *Scotland's Scenic Heritage*, Perth: CCS.

—— (1985) *Leisure in the Scottish Countryside: The Results of a Survey*, Perth: CCS.

—— (1986) *A Park System and Scenic Conservation in Scotland*, Perth: CCS.

—— (1990) *The Mountain Areas of Scotland*, Perth: CCS.

Countryside Review Committee (1976) *The Countryside – Problems and Policies*, HMSO.

Crowe, S. (1956) *Tomorrow's Landscape*, Architectural Press.

—— (1966) *Forestry in the Landscape*, Edinburgh: FC.

Darling, F.F. (1971) *Wilderness and Plenty: The Reith Lectures 1969, Ballantine*.

Dawson, E. (1980) *The Environmental Movement in the 1980s*, CoEnCo.

Department of the Environment (1977) *Nature Conservation and Planning*, Circ.108/77, DoE.

—— (1982) *Code of Guidance for Sites of Special Scientific Interest*, DoE.

—— (1986) *Conservation and Development: The British Approach*, DoE.

—— (1988) *Protecting Your Environment: A guide*, DoE.

—— (1993) *Countryside Survey 1990*, DoE.

—— (1995) *Rural England: A Nation Committed to a Living Countryside*, DoE.

Dobson, A. (1990) *Green Political Thought*, Routledge.

—— and Lucardie, P. (eds) (1995) *The Politics of Nature; Explorations in Green Political Theory*, Routledge.

Dower, J. (1945) *National Parks in England and Wales*, Cmnd 6378, HMSO.

Dunnet, G.M. *et al.*(1986) *Badgers and Bovine Tuberculosis*, HMSO.

Durrell, L. (1986) *State of the Ark*, Gaia Books/Bodley Head.

'Ecologist' (1972) *A Blueprint for Survival*, Penguin.

Edwards, R. (1991) *Fit for the Future*, Cheltenham: CC.

Ellington, A. and Burke, T. (1981) *Europe: Environment. The European Communities' Environmental Policy*, Ecobooks.

English Nature (1993) *Progress '93*, Peterborough: EN.

Environment Council, *habitat* (all issues), Environment Council.

Fairbrother, N. (1972) *New Lives, New Landscapes*, Penguin.

Fisher, J. (1940) *Watching Birds*, Penguin.

Fitter, R. (1963) *Wildlife in Britain*, Penguin.

Fitter, R. and Scott, P. (1978) *The Penitent Butchers*, FFPS.

Forestry Commission (1965) *State Forest Memorandum*, Edinburgh: FC.
—— (1986) *The Forestry Commission and Conservation*, Edinburgh: FC.
Freethy, R. (1981) *The Making of the British Countryside*, Newton Abbot: David and Charles.
Gare, A. (1995) *Postmodernism and the Environmental Crisis*, Routledge.
Garner, J.F. and Jones, B.L. (1987) *Countryside Law*, Shaw.
Gaze, J. (1988) *Figures in a Landscape: A History of the National Trust*, Barrie and Jenkins.
Goldsmith, E. and Hildyard, N. (eds) (1986) *Green Britain or Industrial Wasteland?* Oxford: Polity Press.
Goldsmith, F.B. and Warren, A. (1993) *Conservation in Progress*, Chichester: Wiley.
Goode, D. (1981) *Recent Losses of Wildlife Habitat in Britain*, NCC.
—— (1986) *Wild in London*, Michael Joseph.
Gooders, J. (1967) *Where To Watch Birds*, Andre Deutsch.
Goodman, G.T. *et al.* (1965) *Ecology and the Industrial Society*, Oxford: Blackwell.
Green, B. (1985) *Countryside Conservation*, Allen and Unwin.
Green, B.H. (ed.) (1979) *Wildlife Introductions to Great Britain*, Peterborough: NCC.
Gresswell, P. (1971) *Environment: An Alphabetical Handbook*, John Murray.
Harrison, J. (1973) *A Wealth of Wildfowl*, Corgi.
—— (1974) *The Severnoaks Gravel Pit Reserve*, Chester: WAGBI.
Hill, H. (1980) *Freedom To Roam*, Ashbourne: Moorland Publishing.
Hirsch, F. (1977) *The Social Limits to Growth*, Routledge.
Her Majesty's Stationery Office (1976) *Convention on Wetlands of International Importance especially as Waterfowl Habitat*, Cmnd 6465, HMSO.
—— (1980) *Convention on the Conservation of European Wildlife and Natural Habitats*, Cmnd 7809, HMSO.
—— (1980) *Convention on the Conservation of Migratory Species of Wild Animals*, Cmnd 7888, HMSO.
—— (1984) *Agriculture and the Environment*, HMSO.
—— (1990) *This Common Inheritance: Britain's Environmental Strategy*, Cmnd 1200, HMSO.
—— (1994) *Britain 1995: An Official Handbook*, HMSO.
—— (1994) *Strategy for Sustainable Development*, HMSO.
—— (1994) *Sustainable Forestry – The UK Programme*, HMSO.
—— (1994) *UK Biodiversity Action Plan*, HMSO.
Hobhouse, Sir A. (1947) *Report of the National Park Committee (England and Wales)*, Cmnd 6628, HMSO.
Hoskins, W.G. (1977) *The Making of the English Landscape*, Hodder and Stoughton.
Huxley, Sir J. (1947) *Report of the Committee on Nature Conservation in England and Wales*, Cmnd 7122, HMSO.
Hywel-Davies, J. and Thom, V. (1984) *Guide to Britain's Nature Reserves*, Macmillan.
Irvine, S. and Ponton, A. (1988) *A Green Manifesto*, Optima.
Institute of Terrestrial Ecology (1994) *1993 Hedgerow Survey*, ITE.
International Union for Conservation of Nature and Natural Resources (1980) *A World Conservation Strategy: Living Resource Conservation for Sustainable Development*, Cambridge: IUCN.
IUCN – the World Conservation Union (1991) *Caring for the Earth: A Strategy for Sustainable Living*, Earthscan.
—— (1994) *Parks for Life: Action for Protected Areas in Europe*, Cambridge: IUCN.
—— (1994) *A Guide to the Convention on Biological Diversity*, Cambridge: IUCN.

Jackson, C.E. (1968) *British Names of Birds*, Aylesbury: Witherby.

Johnson, B. (1983) *An Overview: Resourceful Britain*, Kogan Page.

King, A. and Clifford, S. (1985) *Holding Your Ground*, Aldershot: Wildwood House.

Kinnersley, D. (1994) *Coming Clean: The Politics of Water and the Environment*, Penguin.

Leopold, A. (1949) *A Sand County Almanac*, New York: Oxford University Press.

Lowe, P. and Goyder, J. (1983) *Environmental Groups in Politics*, Allen and Unwin.

—— *et al.* (1986) *Countryside Conflicts: the Politics of Farming, Forestry and Conservation*, Aldershot: Gower.

Lowenthal, D. and Binney, M. (eds) (1981) *Our Past Before Us: Why Do We Save It?*, Aldershot: Temple Smith.

Mabey, R. (1980) *The Common Ground*, Hutchinson.

MacEwen, A. and M. (1982) *National Parks: Conservation or Cosmetics*, Allen and Unwin.

MacEwen, M. (ed.) (1976) *Future Landscapes*, Chatto and Windus.

Marren, P. (1995) *The New Naturalists*, HarperCollins.

Matthews, G.V.T. (1993) *The Ramsar Convention on Wetlands: Its History and Development*, Cambridge: IUCN.

Meadows, D. *et al.* (1972) *The Limits to Growth – A Report for the Club of Rome*, Pan.

Mellanby, K. (1967) *Pesticides and Pollution*, Collins.

Mence, A. (ed.) (1981) *IUCN: How It Began; How It Is Growing Up*, Cambridge: IUCN.

Mill, J.S. (1857) *Principles of Political Economy, vol. 2*, John Parker.

Milner, J.E. *et al.* (1990) *The Green Index: A Directory of Environmental Organisations in Britain and Ireland*, Cassell.

Ministry of Agriculture, Fisheries and Food / Department of the Environment (1995) *Environmental Land Management Schemes in England (Consultation Document)*, MAFF/DoE.

Moore, N.W. (1987) *The Bird of Time: The Science and Politics of Nature Conservation*, Cambridge: Cambridge University Press.

—— (1991) *Conservation in the Nineties*, ECOS Conservation Comment.

Moriarty, F. (1975) *Pollutants and Animals*, Allen and Unwin.

Mostyn, B.J. (1979) *Personal Benefits and Satisfactions Derived from Participation in Urban Wildlife Projects: A Qualitative Evaluation*, Peterborough: NCC.

National Rivers Authority (1991) *The Quality of Rivers, Canals and Estuaries in England and Wales*, Bristol: NRA.

National Trust (1986) *Background Information*, NT.

Natural Environment Research Council (1973) *Marine Wildlife Conservation: An Assessment of a Threat to Marine Wildlife and the Need for Conservation Measures*, Peterborough: NERC.

Nature Conservancy Council (1970) *Twenty-one Years of Conservation*, Peterborough: NCC.

—— (1977) *Nature Conservation and Agriculture*, Peterborough: NCC.

—— (1984) *Nature Conservation in Great Britain*, Peterborough: NCC.

—— (1989) *Woods, Trees and Hedges: A Review of Changes in the British Countryside*, Peterborough: NCC.

—— (1990) *Nature Conservation and Agricultural Change*, Peterborough: NCC.

—— (1991) *17th Report*, Peterborough: NCC.

Nature Reserves Investigation Committee (1943) *Nature Conservation in Great Britain*, SPNR.

—— (1945) *National Nature Reserves and Conservation Areas in England and Wales*, SPNR.

Nicholson, M. (1970) *The Environmental Revolution*, Hodder and Stoughton.
Nicholson, M. (1987) *The New Environmental Age*, Cambridge: Cambridge University Press.
O'Riordan, T. (1976) *Environmentalism*, Pion.
—— (1989) 'Nature Conservation under Thatcherism: the legacy and the prospect', *ECOS* 10(4): 4–8 .
Parkes, C. and Thornley, J. (1987) *Fair Game: The Law of Country Sports and the Protection of Wildlife*, Pelham.
Parsons, C. (1982) *True to Nature*, Wellingborough: Patrick Stephens.
Pearce, D.W. (ed.) (1991) *Blueprint 2: Greening the World Economy*, Earthscan.
—— *et al.* (1989) *Blueprint for a Green Economy*, Earthscan.
Pepper, D. (1984) *The Roots of Modern Environmentalism*, Croom Helm.
—— (1993) *Eco-socialism: From Deep Ecology to Social Justice*, Routledge.
Perrin, J. (1988) 'Friends in need', *Environment Now* 3.
Perring, F.H. (1975) 'Problems of conserving the flora of Britain', *BSBI News* 9.
—— (1982) *The Voluntary Nature Conservation Movement in the UK*, Nettleham: RSNC.
—— and Farrell, L. (eds) (1977) *British Red Data Books: 1 Vascular Plants*, Nettleham: SPNC.
Perry, R. (1978) *Wildlife in Britain and Ireland*, Croom Helm.
Peterken, G.F. (1981) *Woodland Conservation and Management*, Chapman and Hall.
Pollard, E. *et al.*(1974) *Hedges*, Collins.
Porchester, Lord (1977) *A Study of Exmoor*, HMSO.
Prentice, R.J. (undated.) *Conserve and Provide: A Brief History of the National Trust for Scotland*, Edinburgh: NTS.
HRH Prince Philip (1978) *The Environment Revolution*, Andre Deutsch.
Purseglove, J. (1988) *Taming the Flood: A Natural History of Rivers and Wetlands*, Oxford: Oxford University Press.
Pye-Smith, C. (1990) *In Search of Neptune: A Celebration of the National Trust's Coastline*, National Trust.
Rackham, O. (1980) *Ancient Woodland: Its History, Vegetation and Uses in England*, Edward Arnold.
—— (1986) *The History of the Countryside*, Dent.
Ramsay, Sir J.D. (1945) *National Parks: A Scottish Survey*, Cmnd 6631, Edinburgh: HMSO.
—— (1947) *National Parks and the Conservation of Nature in Scotland*, Cmnd 7235, Edinburgh: HMSO.
Ratcliffe, D.A. (1970) 'Changes attributable to pesticides in egg breakage frequency and eggshell thickness in some British birds', *Journal of Applied Ecology* 7: 67–115.
—— (ed.) (1977) *A Nature Conservation Review: The Selection of Biological Sites of National Importance to Nature Conservation in Britain*, 2 vols, Cambridge: Cambridge University Press.
—— (1980) *The Peregrine Falcon*, Berkhamsted: Poyser.
—— (1989) 'The Nature Conservancy Council 1979–1989', *ECOS* 10(4), 9–15.
Ritchie, J. (1949) *Final Report on Nature Reserves in Scotland*, Edinburgh: HMSO.
Rook, D. (1969) *Protecting Britain's Birds*, Sandy: RSPB.
Rose, C. and Secrett, C. (1982) *Cash or crisis: the imminent failure of the Wildlife and Countryside Act*, BANC/FoE.
Rose, F. (1984) 'Reserves for everyone?', *Natural World* 10.
Rothschild, M. (1979) *Nathaniel Charles Rothschild 1877–1923*, Cambridge: Cambridge University Press.
—— (1992) 'Road sense', *BBC Wildlife* 10(7): 30–2 .

Rowe, A. (1992) 'Green government gets off the ground', *BBC Wildlife* 10(4): 54–6.

Royal Society for the Protection of Birds (1995) 'Conservation lobby forces change in Bill', *Birds* 16(1): 5.

Ryle, G. (1969) *Forest Service: The First Forty-five Years of the Forestry Commission of Great Britain*, Newton Abbot: David and Charles.

Samstag, T. (1989) *For the Love of Birds: The Story of the RSPB*, Sandy: RSPB.

Sandford, Lord (1974) *Report of the National Park Policy Review Committee*, HMSO.

Schama, S. (1995) *Landscape and Memory*, HarperCollins.

Scott, Mr Justice (1942) *Report of the Committee on Land Utilisation in Rural Areas*, Cmnd 6378, HMSO.

Scottish Natural Heritage (1993) *Sustainable Development and the Natural Heritage: The SNH approach*, Perth: SNH.

—— (1994) *Third Operational Plan – Review of 1993–94 and Work Programme 1994–95*, Perth: SNH.

—— (1995) *Enjoying the Outdoors: A Programme for Action*, Perth: SNH.

Scottish Wildlife Trust (1995) *Annual Review 1994–1995*, Edinburgh: SWT.

Sheail, J. (1976) *Nature In Trust: The History of Nature Conservation in Britain*, Glasgow: Blackie.

—— (1981) *Rural Conservation in Inter-war Britain*, Oxford: Oxford University Press.

—— (1985) *Pesticides and Nature Conservation: The British Experience 1950–75*, Oxford: Oxford University Press.

—— (1987) *Seventy-five Years in Ecology: British Ecological Society*, Oxford: Blackwell.

Shell Better Britain Campaign (1995) *Interactive*, Birmingham: Shell Better Britain Campaign.

Shoard, M. (1980) *The Theft of the Countryside*, Aldershot: Temple Smith.

—— (1987) *This Land Is Our Land: Struggle for Britain's Countryside*, Grafton.

—— (1988) 'Lie of the land', *Environment Now* 10.

Simmons, S.A. *et al.* (1990) *Nature Conservation in Towns and Cities: A Framework for Action*, Peterborough: NCC.

Smout, C. (1993) *The Highlands and the Roots of Green Consciousness 1750–1990*, Perth: SNH.

Society for the Protection of Nature Reserves (1971) 'Farming and the Soil', *Conservation Review* 3, Alford: SPNR.

Sommerville, A. (1995) 'New life from Brussels?' *Scottish Wildlife* 26: 14–15.

Stamp, D. (1969) *Nature Conservation In Britain*, Collins.

Stamp, L.D. and Hoskins, W.G. (1963) *The Common Lands of England and Wales*, Collins.

Standing Commission on National Parks (1938) *The Case for National Parks*, SCNP.

Steele, R.C. (1972) *Wildlife Conservation in Woodlands*, Edinburgh: FC.

Stephenson, T. (1989) *Forbidden Land: the Struggle for Access to Mountain and Moorland*, (ed. A. Holt) Manchester: Manchester University Press.

Stevenson, D. (1972) *50 Million Volunteers: A Report on the Role of Voluntary Organisations and Youth in the Environment*, HMSO.

Tansley, A.G. (1939) *The British Islands and their Vegetation*, Cambridge: Cambridge University Press.

—— (1945) *Our Heritage of Wild Nature: a Plea for Organised Nature Conservation*, Cambridge: Cambridge University Press.

Taylor, G.R. (1970) *The Domesday Book*, Thames and Hudson.

Thomas, K. (1983) *Man and The Natural World*, Allen Lane.

United Nations Conference on Environment and Development (1993) *The Earth Summit: The United Nations Conference on Environment and Development*, Graham and Trotman.

Vesey-Fitzgerald, B. (1969) *The Vanishing Wild Life of Britain*, MacGibbon and Kee.

Wall, D. (ed.) (1994) *Green History: A Reader in Environmental Literature, Philosophy and Politics*, Routledge.

Warren, A. and Goldsmith, F.B. (1974) *Conservation in Practice*, Wiley.

—— (1983) *Conservation in Perspective*, Chichester: Wiley.

Waterson, M. (1994) *The National Trust: The First Hundred Years*, BBC/NT.

Westmacott, R. and Worthington, T. (1976) *New Agricultural Landscapes*, Cheltenham: Countryside Commission.

White, G. (1789) *Natural History of Selborne*, Benjamin White.

Williams, R. (1973) *The Country and the City*, Chatto and Windus.

Worcester, R. and Smith, C. (1994) 'Growing greens', *Earth Matters* 24: 8–9.

World Commission on Environment and Development (1987) *Our Common Future* (the Brundtland report), Oxford: Oxford University Press.

World Wide Fund for Nature (1991) 'Caring for the Earth', *WWF 1991 Review*: 6–7.

World Wide Fund for Nature (1995) 'Ariel washes greener', *WWF News* Spring: 14.

Wright, F.J. *et al.*(eds) (1993) *Action for Biodiversity in the UK*, Peterborough: JNCC.

In addition, the annual reports and reviews of the major official agencies and voluntary bodies are a vital source of information and comment.

Species index

PLANTS

General index